TWENTY WORLDS

● UNIVERSE

SERIES EDITOR: James Geach

Titles in the Series
Twenty Worlds: The Extraordinary Story of Planets Around Other Stars NIALL DEACON

Forthcoming
Simulating the Cosmos ROMEEL DAVÉ
Telescopes / Observing the Universe SARAH KENDREW

TWENTY

THE EXTRAORDINARY STORY OF
PLANETS AROUND OTHER STARS

WORLDS

NIALL DEACON

REAKTION BOOKS

Published by
REAKTION BOOKS LTD
Unit 32, Waterside
44–48 Wharf Road
London N1 7UX, UK
www.reaktionbooks.co.uk

First published 2020
Copyright © Niall Deacon 2020

Printed and bound in China
by 1010 Printing International Ltd

A catalogue record for this book is available from the British Library

ISBN 978 1 78914 338 6

CONTENTS

PREFACE

Thirty years ago there were eight worlds: our own Earth, five planets known since antiquity and another two discovered after the invention of the telescope. Astronomers were observing subtle fluctuations in radiation left over from the Big Bang, but didn't know if there were planets in orbit around stars neighbouring the Sun.

Now we know thousands of worlds.

In this book you will find twenty worlds, each a planet, some similar to the Earth, some very different. Each has its own story to tell, something important both for it and for many other worlds. Together their stories show the diversity of worlds around other stars, the range of techniques to find and characterize them and the physical processes that affect their formation, their likelihood to host life and ultimately their demise.

In the Appendix you will find the characteristics of each of the twenty worlds and in the References details of the wonderful studies that identified and characterized these planets. There you will also see the names of some the amazing scientists who worked on these studies, but bear in mind that the listed names are often limited to team leaders and that all astronomy is built on the work of others, be they collaborators, other scientists who did previous work on a topic, engineers building the instruments for observing the universe or the myriad of observatory and university staff from telescope operators to administrators to cleaners.

My hope is that this book shows how we know things about planets. Science isn't magic, an arcane lore of quantum theory and general relativity that a few chosen wizards can manipulate. It is, put simply, a collection of ways of working things out. It could be working out if there is a planet in orbit around a star, working out how massive that planet is or working out what the best model is for its atmosphere. If this book does at some point start to seem like bamboozling sorcery then that is my fault for not explaining things clearly.

Finally, I hope this book gives a good overview of our knowledge of a diverse and ever-evolving area of astronomy. I also hope it gives a guide to how astronomers have learned so much about this topic in less than thirty years.

INTRODUCTION:
AN ORDERED FAMILY PORTRAIT

You might want to get out the extra blanket tonight.

This is the suggestion given by weather forecasters in Hawai'i any time the night-time low drops below 20°C (68°F). That is to be expected in a place where the weather is generally so pleasant and predictable that the nightly news has time to include surf, volcanic gas and jellyfish forecasts.

The Hawai'i weather forecast portrays a stable, almost ideal climate for humans to live in. Compared to the freezing, wind-swept Antarctic plains or the scorched deserts of Death Valley, Honolulu provides an environment for humans that is not too hot (rarely getting above 30°C (86°F)) and not too cold (occasionally dipping below 20°C). This range of temperatures seems staggeringly small in the context of a bracing −81°C (−114°F) at night at Lake Vostok or a blistering 57°C (135°F) during the day in rural California. A night in the Antarctic would probably need a bit more than the extra blanket recommended by the weather presenter on TV in Hawai'i. But these plunging temperatures and the parched desert of Death Valley are a tranquil archipelago compared to the huge variety of planets seen around other stars.

On any particular day the temperature in Honolulu will vary by about 10°C (18°F), one-fourteenth of the range seen over the Earth in the last few hundred years of meteorological measurements. Our planet is, however, a relatively cushy and tranquil place to live. The solar system sees variations from −240°C (−400°F) in the frozen icy bodies that inhabit its outer reaches, such as Pluto, to the wilting heat of 470°C (880°F) in the thick atmosphere of Venus. Even this 710°C (1,290°F) range is small compared to what we see in planets around other stars. These extrasolar planets (or exoplanets) are commonly seen with temperatures over 1,000°C (1,832°F), with the current record being 4,300°C (7,772°F) in the roasting atmosphere of KELT 9b orbiting close to its white-hot parent star.[1] The range of temperatures seen on planets around other stars is roughly thirty times the range of temperatures seen on the Earth.

The temperature range doesn't cover the full diversity of planets. There are airless, desolate surfaces, worlds shrouded in choking toxic gas and even clouds of rubies and sapphires and planets made of diamond. There's another world of worlds out there.

PLANETS STOOD OUT to many early societies because they moved through the heavens. In ancient Mesopotamia planets were 'wild sheep' ambling across the sky.[2] To the Greeks they were 'wandering stars', moving through the heavens relative to the 'fixed stars'. In fact the Greek word to wander, *planētēs*, gives us the word 'planet'.[3] For much of the history of astronomy those who practised it doubled as court priests or astrologers or as the keepers of the calendar. The first appearance of Sirius marked the onset of the crucial Nile floods in Egypt and the rising of the Pleiades star cluster in the East opposite the Sun setting in the West marked the beginning of the Hawaiian festival season of Makahiki. Thus it was crucial for agriculture, ritual and the operation of early states that positions of stars were calculated accurately.

Predicting planetary paths across the night sky requires a model of the cosmos. The map of the heavens used by most prominent ancient Greek thinkers placed the Earth at the centre with the planets and Sun orbiting around it. This relied on a series of complex celestial motions to replicate the movements of the planets. The third-century BCE Greek astronomer Aristarchus of Samos found that a model with the Sun in the centre of the known universe fitted fairly well. While the idea died out in the West for more than a millennium, it was independently developed by astronomers in the medieval Islamic world and in the Indian Kerala school of astronomers. What Aristarchus also stated was something more radical that had rarely been expressed in Greek philosophy before, that the 'fixed stars', those points of light that stayed put while the planets wandered, were other suns.[4] This, while not widely adopted, was an early speculation on the nature of stars and planets as physical bodies rather than mere light sources to be observed.

Writing in the first century BCE, the Chinese astronomer Jing Fang noted, 'the Moon and the planets are Yin, they have shape but no light.'[5] This was a clear statement that planets do not have a light source of their own but are bodies that reflect the Sun's light. In Europe, the Moon and the planets were considered by Aristotle and the later Church figures who adopted his doctrine to be flawless heavenly bodies, perfect spheres. That was until someone took a closer look.

Galileo Galilei was so brilliant they almost named him twice. He was the first person to publish widely read telescopic observations of the Sun, Moon and planets. These blew away the idea that the Sun and the Moon were perfect creations. The Sun was spotty and the Moon was scarred with jagged mountains. Saturn had a strange elongated shape, almost as if it had two handles. And then there was Jupiter, which appeared to have four stars orbiting around it. These were not ideal, flawless heavenly objects; these were real, often strange physical bodies with imperfections and properties like our Earth.

BESIDES EARTH, there are seven major worlds in the solar system: Mercury, Venus, Mars, Jupiter and Saturn can all be seen with the naked eye, while Uranus and Neptune were discovered using telescopes. These planets fit into three families, the rocky inner terrestrial planets (Mercury, Venus, Earth and Mars), the two large gas giants (Jupiter and Saturn) and the two outer ice giants (Uranus and Neptune).

The four inner planets seem to be very different; scorched Mercury, Venus with its choking toxic atmosphere, our own temperate Earth and dry and barren Mars. However, they are alike in one important way: they mostly consist of rock, something that is made from heavier elements such as silicon, oxygen and iron. The latter three of these planets each have a small layer of atmosphere around them composed of heavier gases such as nitrogen, oxygen and carbon dioxide with almost no hydrogen and helium. The terrestrial planets have relatively weak gravitational pulls, making it hard for them to hang on to any lighter hydrogen and helium in their atmosphere. The one exception is Mercury, a planet whose extremely thin atmosphere has a small amount of these gases that is constantly replenished by the hydrogen and helium that is blasted out of our central star as part of the solar wind.

Jupiter and Saturn dominate our planetary system, with Jupiter containing more material than the other planets put together and the two worlds combined making up more than 90 per cent of the solar system's planetary mass. Both probably have rocky cores, with Saturn's core being about nine to 22 times the mass of the Earth.[6] In the early solar system planetary embryos called protoplanets competed for the gas and solid material left over from the formation of the Sun. Larger protoplanets have greater gravitational pulls and can thus suck up more material. The larger embryos of Jupiter and Saturn could thus draw towards them huge amounts of the abundant hydrogen and helium in the early solar system, allowing them to reign over the lower-mass planets. Having larger masses than the terrestrial planets, Jupiter and Saturn were able to hold on to this gas due to their greater gravitational pulls.

Below the cloud bands and storms of Jupiter's atmosphere lies a strange ocean made not of water, but of hydrogen and helium. The atmosphere above is so massive that its weight pushes down on the material below with such pressure that it is forced into liquid form. Below this is yet another ocean made of an extreme form of the same elements. At this deeper point the pressure is so huge that the hydrogen atoms are stripped of their electrons and the ocean begins to behave like liquid metal. You may think

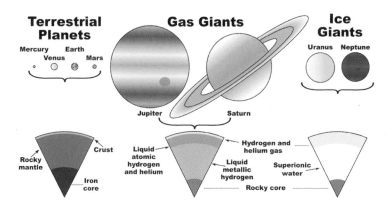

Our solar system's ordered family portrait. The planet sizes are to scale but they are shown much closer together than they actually are. The Sun is about ten times Jupiter's radius.

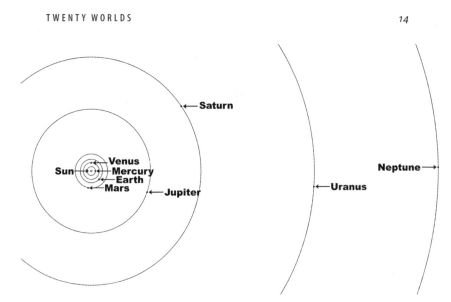

The orbits of the planets to scale. The Sun is twenty times its true size. The dots representing the planets are not to scale. The planet orbits are not the simple circles shown but are slightly elliptical.

of the element mercury as the most obvious example of a liquid metal. Liquid metallic hydrogen would look similar, reflecting light, but with a substantially lower viscosity than mercury, making it runny like water.

The last of our three categories of major planets, ice giants, lack the neatly explained origins story of the gas giants. They lie far from the Sun, in regions where planet formation would have been more difficult. It is thought they might have formed closer to Jupiter and Saturn and been kicked out to wider orbits by their heavier, greedy cousins. Uranus and Neptune are both quite different from the gas giants and the terrestrial planets. They have rocky cores but were not massive enough to collect large amounts of hydrogen and helium in the early solar system, unlike the gas giants. They do, however, have enough of these two elements that they dominate the planets' atmospheres. Ice giants also have an intermediate layer between the core and the atmosphere made

up of a strange high-pressure form of water, mixed with ammonia and methane. This superionic water is both solid and liquid and is often referred to as ice, but is very different in structure from the solid water you would see on a frosty day on Earth.[7] The 'ice' in 'ice giants' refers not to the current state of the water, ammonia and methane in Uranus and Neptune but to the source of these chemicals. When both planets were forming early in the solar system's history, they sucked up huge quantities of solid icy material in the frigid regions far from the Sun.

The planets in our solar system seem easy to characterize: rocky inner planets, giants that are made mostly of hydrogen and helium, and planets with a large amount of water and an outer atmosphere of lighter gases.

CARAVAGGIO'S PAINTING *The Taking of Christ* hangs in the National Gallery of Ireland.[8] Lost for more than two hundred years until its rediscovery in 1990, it shows Christ being detained by Roman soldiers after being betrayed by Judas Iscariot. Caravaggio, a man who would spend the last four years of his life on the run for murder, included himself in the painting as St Peter. As an artist Caravaggio is known for his dark shadows and bold character illumination. It is one such illumination that grabs the eye in this painting, a glint of light off the soldier's arm thrusting towards Jesus' neck. But this is not the arm of a Roman legionary, it's the arm of a seventeenth-century soldier. Rather than attempt to re-create the period costume of these biblical antagonists, Caravaggio dresses Christ's assailants as troops from his own time. This sort of anachronism, the insertion of contemporary characters into biblical or mythological scenes, is common in Western art.

It is not just the view of the past that is susceptible to such anachronism. Around the turn of the twentieth century, a number of French artists produced images of what they thought the year 2000 would be like.[9] The images show somewhat steampunk labour-saving devices and those darlings of futurology, personal flying machines. In some cases they demonstrate surprising foresight, the envisioned 'battlecars', for example, fulfil the same function as modern tanks. In the school classroom books are fed into a giant mincing machine before being transmitted to students via a headset, a rather clunky foreshadowing of our information society. In all cases the technology used takes forms familiar in the late nineteenth century. There are no computers, rockets or nuclear weapons, only drive belts, propellers and Gatling guns. Yet again the artists' knowledge of their own familiar present is applied to a far-flung time frame.

The same thing is easy to do when thinking of far-flung corners of the universe, to extrapolate from the familiar. This is perfectly natural, as we see the planets in our solar system have some order: rocky planets in orbits of months to a few years, gas giants with orbits of a few tens of years and ice giants in orbits with periods of about a hundred years. You would assume that other planetary systems would follow similar patterns. But the first planets found around other Sun-like stars were to prove these assumptions very, very wrong indeed.

ALIEN WORLDS

1

A WORLD BEYOND EXPECTATION

Edinburgh of the Seven Seas sits on the volcanic island of Tristan da Cunha in the South Atlantic. A few hundred metres across, on an island less than 12 kilometres (7.5 mi.) wide and with a population of a couple of hundred, it is the most remote permanent settlement on Earth, over 2,400 kilometres (1,500 mi.) from the nearest town on the island of St Helena.

Our solar system is pretty isolated too. Let's scale the solar system to be the size of Edinburgh of the Seven Seas. The Sun would be in the centre with the terrestrial planets around it, Jupiter about 45 metres (150 ft) away at the Albatross bar, distant Neptune 270 metres (885 ft) away on the edge of town on the way to the docks. In this scenario Proxima Centauri, the nearest star to the Sun, sits as far away from the town centre as St Helena does from Tristan da Cunha.

EACH OF THE few hundred billion stars in the Galaxy could, like our Sun, have its own little planetary town huddled around it. But in 1990 all we could say was that one star, the Sun, out of hundreds of billions, definitely hosted planets.

We have seen that huge distances would separate any little planetary towns across the vast oceanic expanse of interstellar space. This means that from a distance planets will appear to be very close to their star, making it difficult to resolve the planets from the star. Stars are also much brighter than planets. (In visible light Jupiter is more than 200 million times fainter than the Sun.) These two facts combined mean that it is difficult to directly image a planet around another star. To make finding planets easier, astronomers need to try a few clever tricks, one of which involves a cosmic balancing act.

A see-saw is simple enough to understand: if someone lighter than you sits on the other end from you then you thump down on the ground; conversely, if your see-saw partner is heavier than you, you shoot up in the air and are left high up with your legs dangling. Adjusting the pivot point of the see-saw can alter this balancing act. By moving the point where the see-saw pivots closer to the heavier of the two ends it is possible to balance the see-saw. It does not matter who is on the other end of the apparatus from you; a child, an adult, a troupe of acrobats or an escaped rhinoceros from the local zoo, there will always be a special pivot point you can choose on the see-saw to balance the weight. The bigger the difference in weight between you and whatever is on the other end of the see-saw, the more the pivot point has to be moved from the midpoint of the see-saw. Take the example of sitting across from a rhino. In that case the pivot point would need to be thirty times closer to the harrumphing herbivore than it was to you.

You might think that the Sun is the immovable centre of the solar system, but it moves. The Earth and the Sun are a bit like you and that rhinoceros on the see-saw. There is a balance point between the two, a special pivot point called the centre of mass, which can be chosen so that if the Sun is on one side of a see-saw and the Earth at the other the two would be balanced. There's not some huge interplanetary see-saw in the solar system, but there is a centre of mass between the Sun and the Earth. It is not the centre of the Sun that we orbit but this pivot point situated about 3 millionths of an Earth–Sun distance away from the centre of the Sun. The Sun's radius is about half of 1 per cent of an Earth–Sun distance, so the pivot point the Earth orbits is still inside the Sun. The strange thing is, the Sun itself also orbits this pivot point.

The little orbit the Sun makes around this pivot point is tiny because the Earth is so small compared to the Sun (3 millionths of its mass). That is like you being on the other end of the see-saw from 10,000 rhinos, more than one-third of the current global rhinoceros population. But consider Jupiter, more than three hundred times the Earth's mass and five Earth–Sun distances from the Sun. The balance point between Jupiter and the Sun is just outside the Sun. Both Jupiter and the Sun orbit around this point: Jupiter moving the Sun around in a little orbit; the Sun moving Jupiter around in a much bigger one.

So planets can move their stars around. Why does that matter? Because this leaves a mark on the light we receive from a star. To find out how, we need to head over to the starting grid where the cars are revving up their engines.

THE PING OF a drive off the tee in golf, the sound of leather on willow at a cricket match, some sports have sounds that define them. For Formula One racing that sound is 'neeeeeeeoooooooowww' as a car screams past you. This is a sound that hides an interesting bit of physics.

A Formula One car doesn't change the tone of its engine noise when it is going at a constant speed on a straight. But when it comes towards you it sounds higher than once it has gone past and is heading away from you. Sound is a wave moving through the air. When the car is coming towards you the sound waves get squashed together, giving the engine noise a higher tone. When the car moves away from you the opposite happens and the waves are stretched out, giving the engine a lower tone.

Astronomers see a similar shift in starlight. When a star is moving towards the Earth its light waves are squashed up, making them bluer. When the star moves away from us the opposite happens and the star's light becomes redder.

Light comes in many colours. Astronomers split light up into its spectrum of colours to study stars and galaxies in more detail. Typically, they refer to these colours by the wavelength of the light: the redder the light, the longer the wavelength.

The light a star gives off does not come from a solid or liquid surface. A star is a big ball of plasma (ionized gas) with a hot, dense core in the middle containing a nuclear furnace. Here the

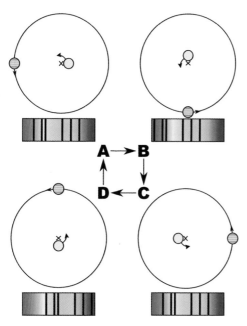

This not-to-scale diagram shows the orbit of a planet and star around the centre of mass of their system. The star's spectrum seen by an observer on Earth (here positioned on the right and represented by an eye) shows the pattern of dark lines moving from red to blue then back again.

temperature and pressure are so high that hydrogen can fuse into helium, producing energy. The heat from this furnace slowly seeps out to the outer layers of the star, keeping them hot. The material inside the star is hot so it glows, emitting light. But the material is also very dense so the aforementioned light travels only a short distance before hitting or being deflected by one of the ions in the plasma. The further away you go from the star's centre, however, the cooler and less dense the plasma becomes. Eventually you get to the point where the density of the plasma falls to the level where light can escape from the star into outer space. Given this is the bit of the star we see, as it is where the light comes from, you could consider it as being like the surface of the star. But before the light from this 'surface' gets out into the cosmos, it must pass through gas in the star's upper atmosphere. This gas is cooler still and may even have atoms in it. These atoms are fussy feeders that leave telltale fingerprints on a star's spectrum.

DON'T STAND BEHIND me in the queue for a buffet. I'm not the pickiest eater but I will spend forever carefully teasing the lettuce and tomatoes out of the salad bowl and avoiding the cucumber. Atoms are even worse than me when it comes to being selective eaters.

The atoms in a star's atmosphere are finicky feasters on its spectrum. They nibble away, absorbing light at very specific wavelengths, creating little dark lines in the star's spectrum that we observe. So why do they do this?

An atom is made up of a nucleus surrounded by electrons. It is how these electrons are arranged that is important for an atom's effect on light.

An atom is a bit like an amphitheatre. The nucleus is on the stage in the middle with the electrons sitting on different levels of seating around it. Each of these levels has a different energy. If a photon (a particle of light) comes along with just the right energy, it can kick an electron sitting in one row of seats up to another level. To do this, the photon's energy must be the difference in energies between the lower level of seating and the upper level. The photon's energy must be exactly the difference between the energies of these two levels for this to happen. Similarly, if an electron jumps down a few rows, a photon with a very specific energy is emitted.

It is this specific set of energies that mean that atoms are picky eaters. The energy of a photon is related to its wavelength, so the differences in energies between levels of seating in the atom determine what wavelengths that atom will absorb at. Different elements absorb at different sets of wavelengths, so examining the dark lines on a spectrum will tell you the types of atoms that are eating away at a star's light.

Light comes up from within the star and the atoms in its atmosphere absorb it at certain wavelengths. The electrons that have jumped up a few levels will eventually jump back down, emitting light at those specific wavelengths. The re-emitted light, however, will come out in all different directions with much of it going back into the star. Hence the net effect is that the atoms in the star's atmosphere nibble away and produce a series of dark lines in the star's spectrum. It is these dark lines that tell us something about the star's motion, and possibly reveal planets in orbit around it. When a star is moving towards us its light is squashed, moving these dark lines to bluer wavelengths. When a star is moving away from us its dark lines shift to redder wavelengths.

LET US MEET the star that yielded this chapter's planet. It is called 51 Peg and is a yellow star, a little cooler than the Sun, situated fifty light years away from us. For astronomers interested in looking for planets around stars similar to the Sun, 51 Peg was a natural target.

Using a telescope in Haute-Provence in France, astronomers took a series of spectra of 51 Peg, splitting the star's light into its different colours.[1] They also took a spectrum of light from a special lamp in which the metal thorium is heated up to the point where the electrons in its atoms are pushed up to higher levels of seating.[2] They then jump back down closer to the stage, emitting light at specific wavelengths.

This lamp, emitting at known, predictable wavelengths, allowed the astronomers to calibrate their instrument precisely and hence measure accurately the wavelengths of the dark lines in 51 Peg's spectrum.

What did astronomers see? First 51 Peg moved towards the Earth, then away. The dark lines in its spectrum shifted a bit bluer, then a bit redder. Then back to bluer, and then redder again. Using our earlier comparison with a Formula One car, this would be like sitting at the finishing line of a track and, with your eyes closed, listening to a car driving round. You would hear it screaming past you on the home straight, looping round the far end of the circuit and then back past the finish line.

The changes in the wavelengths of the dark lines in the spectrum of 51 Peg showed astronomers this star was moving towards the Earth, then away from the Earth and back again every four days. It was tracing out a tiny orbit as if it was, along with a planet, orbiting a pivot point.

But this planet, named 51 Peg b, was odd. It was pulling the star around so much that it must be a massive planet, at least half as massive as Jupiter. But it was orbiting its star seven times closer than Mercury orbits the Sun.

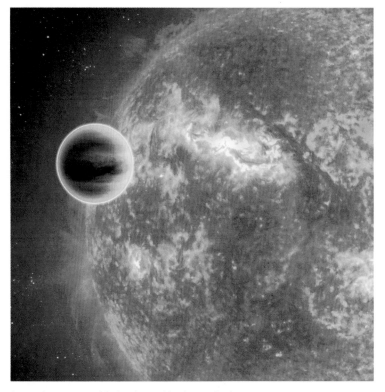

An artist's impression of the hot, Jupiter-like planet 51 Peg b.

This planet, found in 1995 and the first world confirmed to be orbiting another Sun-like star, is truly an alien place, a world beyond expectation. It is so massive that it must be a gas giant like Jupiter. However it is not orbiting far away from its star like the gas giants of the solar system, instead it has a tighter orbit than any solar system planet.

How can we find out more about a world like this? Let's look at a similar planet, one we know a bit more about thanks to sheer luck.

2

A WORLD OBSCURING

It was 11 August 1999, in the days when insurance companies still kept all their policy records on paper. As a student working during the interim between secondary school and university, every morning I was given four hundred letters at 8 a.m. and had until 4 p.m. to file them in policy envelopes somewhere in a vast warehouse. On this particular day I was rushing, trying to get bundles of correspondence filed as quickly as I could to make time to sneak away. It was colder outside than I was expecting and the reason for this soon became clear. Half the Sun appeared to be missing.

In the last chapter we met 51 Peg b, a planet that was detected because it induced subtle changes in the dark lines in its parent star's spectrum. This planet was discovered in 1995 and in the next few years a steady flow of Jupiter-like planets in short orbits (known as hot Jupiters) were found around other stars: 16 Cygni Bb, 47 Ursae Majoris b, Tau Boötis b, Gliese 876 b, HD 168443 b – this list seems like a bewildering array of letters and numbers.

With the exception of a handful of bright stars with special (often Arabic) names or individual interesting stars given the name of a person who studied them, most stars are named with catalogue numbers or letters. This might be a number or letter followed by the name of the constellation the star is found in, or simply a few letters indicating the star catalogue the name comes from, followed by an identification number. Some stars have a few letters followed by some numbers and a plus or minus sign, then some more numbers. These are from surveys that imaged large parts of the sky with the numbers encoding each star's position on the sky. The + and – indicate if the star is in the northern or southern hemisphere, respectively.

Stars can have more than one name. This can be because they are found in more than one catalogue or survey or (for bright stars) because they were given many different names by the great wealth of cultures that have practised astronomy.

After the name of a planet-hosting star you will always see a lowercase letter. If a planet is the first one discovered around a star then it will have the letter 'b', the second planet will be 'c' and so on. This convention comes from stellar multiples, systems where two or more stars orbit each other. If you have two stars orbiting each other then the brighter one will be called 'A' and the fainter one 'B'. Note there is never a planet 'a' in a system. Think of a star name as the family name and the letters after it as the given name. The family name comes first, as in much of East Asia and Hungary. So, for example, 16 Cygni Bb is the first planet found around the second brightest star in a multiple star system in the constellation of Cygnus.

One of the planets found shortly after 51 Peg b was HD 209458 b, a planet with a minimum mass of 0.69 times the mass of Jupiter in a three-and-a-half-day orbit around a star similar to the Sun.[1] You will notice for both 51 Peg b and this planet I have referred to the minimum mass. That is because of a subtle effect related to what a planetary system looks like from Earth.

Astronomers found 51 Peg b and HD 209458 b by identifying that their parent stars were moving towards the Earth, then away from the Earth as the planets orbited. There are a few things we can determine from these measurements. First, we can see how long it takes for them to complete the pattern of coming towards us, away from us and back again. This gives us the orbital period. Second, there is the maximum velocity with which the star is moving towards or away from us. To understand the meaning of this we need to go back to the playground and our friend the rhinoceros.

Previously you were sitting on the see-saw with the rhino. The see-saw was balanced when the pivot point was thirty times further away from you than it was from the rhinoceros. Now let us say a friend of yours decides that playing with this bulky beast looks like fun and jumps on your side of the see-saw. The pivot point of the see-saw must now be adjusted so that it is fifteen times further away from you than it is from the rhino.

The same is true for planetary systems: bigger planets move stars around in bigger orbits than smaller planets do. Planets in closer-in orbits move their stars around in shorter times than planets in wider orbits. Velocity is distance divided by time. The two factors combined (after a little algebraic juggling) mean that massive planets in short orbits move their stars about with the highest velocities. Bigger velocities move the pattern of lines in a star's spectrum further towards the blue or the red. These velocity shifts are easier for astronomers to detect around bright stars than around fainter stars. They are also easier to detect around stars like the Sun that have lots of lines in their spectra. Hence these worlds were the first discovered around brighter Sun-like stars.

This simple picture has one complication. We can only use the shift in the dark lines in a star's spectrum to measure how fast it is going towards or away from us. Imagine a planetary system that was face-on to us, like a fried egg splatted against a wall with the star in the middle of the yolk and the planet going around the edge of the white. The star would trace out a tiny circular orbit round the centre of mass (the star–planet pivot point): up, 12 o'clock, then 3 o'clock, then 6, then 9 then back to 12. What the star would not do is move back and forth in your line of sight. Hence you would not see movement in the dark lines in the star's spectrum. The same system viewed edge-on would see the star move towards us and away from us, as well as left and right. Such a system would be detected by the shift in the lines in the star's spectrum.

The angle at which we see a star–planet system determines how much of a shift we see in the spectral lines of the star. The other things that affect this shift are the mass of the planet and how long it takes to complete its orbit. We know how long it takes to complete an orbit by how long the pattern of towards and away from us takes to complete, so we can remove that factor. But we are still left with the planet mass and the angle we see the system at affecting the size of the spectral line shift. This is a pretty tricky problem; how can we get around it? Well, luckily HD 209458 b does a surprising thing that gives us a lot more information.

In September 1999, by coincidence just after the solar eclipse I witnessed, astronomers were monitoring the brightness of HD 209458 because it had recently been found to host HD 209458 b.[2] As a star similar to the Sun it is quite reserved and mundane, not prone to huge changes in brightness. Soon after starting their observations astronomers saw something strange happen: HD 209458 started to dim. Unlike that solar eclipse where I felt a chill as the Moon covered half the Sun, this dimming was tiny, 1 or 2 per cent of the star's brightness. But the reason for this happening was the same; something was blocking out the light from HD 209458. The same thing occurred a week later when astronomers were at the telescope again. One week is twice the orbital period of HD 209458 b. Both times they had caught the planet partially eclipsing (also known as transiting) its star.

These observations tell us a few things. First, we know that the planet passes between us and the star. This means its orbit must be close to edge-on when seen from the Earth. If a planetary system were face-on, like a fried egg stuck to a wall with the planet going around the outside, then that planet would never come between us and the star. The planet will only obstruct light from its star that is headed to Earth if we are viewing it close to edge-on.

This gives us the last bit of information we need to determine the mass of the planet. Combining an edge-on orbit with the measured velocity from the star's spectral lines and the orbital period tells us that HD 209458 b has a mass of about 69 per cent the mass of Jupiter.

WE KNOW BY HOW much HD 209458 dimmed when its planet transited: between 1 and 2 per cent. The fraction of the star's light that HD 209458 b blocked is called the transit depth. Think back to the solar eclipse I saw in August 1999. Outside the warehouse in Edinburgh I noticed it got a bit colder and a bit darker, but in the very south of England the sky went almost completely dark. That was because the Moon covered more of the Sun when seen from Cornwall. This shows the amount of dimming from a transit or eclipse is related to how much of the bright object the obscuring body blocks.

The Sun is four hundred times further away from us than the Moon and is also four hundred times bigger. This means that the Sun and the Moon appear to be roughly the same size when viewed from Earth, so the Moon can block out the Sun. HD 209458 and HD 209458 b are both roughly the same distance from the Earth, so the amount of dimming is only determined by the relative sizes of the two objects. The transit depth is simply the square of the ratio of the planet's radius to the star's radius. This led astronomers to measure that HD 209458 b had a radius 39 per cent bigger than Jupiter.[3] That is a problem.

Using complex models of the interiors of giant planets, astronomers had expected every object from those that are a bit less massive than Jupiter to possibly eighty times Jupiter's mass to have roughly the same radius as Jupiter. The idea that HD 209458 b has a radius 39 per cent bigger than Jupiter's showed something was missing from the models.

More transiting hot Jupiters were discovered in the years after the observations of HD 209458 b and most showed expanded sizes. One trend astronomers noticed was that the more radiation a hot Jupiter received from its host star, the more inflated it was. So somehow the radiation from the star was heating up the interior of the planet and causing it to swell. Many models have been proposed to explain these puffed-up planets, but one is currently thought to be the best guess, one similar to a process that you might find happening in your home.

Electric bar heaters are found in many households. They could be small portable units used to provide heat around the house or fitted into a wall, perhaps with a fake coal effect similar to the electric fire in my parents' living room when I was growing up. To function, a current flows through a heating element. A similar effect happens in the interior of HD 209458 b and it all starts with the direction the planet faces.

HD 209458 b is a bit like the Moon. No, it's not rocky and it would be a considerably worse place to play golf, but it does (like the Moon) always present the same face to its parent body. In the case of the Moon this is because the Earth's gravity slightly stretches our celestial companion so it has a bulge facing the Earth. As the Moon orbits the Earth, the Earth's gravity is always pulling this bulge towards the Earth. If the Moon were to rotate

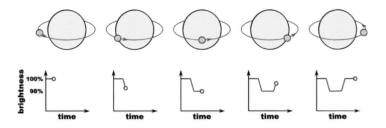

HD 209458 b transits in front of its parent star. The graph beneath shows the reduction in brightness as the star's light is blocked.

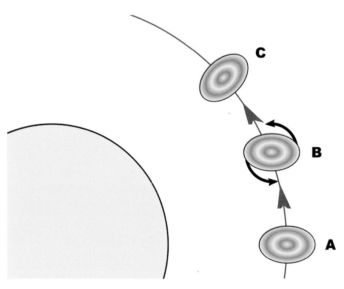

A) A hot Jupiter is stretched by its star's gravity into an oblong shape. B) As the planet moves in its orbit the star's gravity pulls the bulge so it faces the star. C) The planet's bulge is again facing the star. The scale of the deformation of the planet is exaggerated.

slightly, the bulge would be pulled back to face the Earth. This means that we only ever see the same familiar face of the Moon from Earth.

This process, known as tidal locking, also applies to HD 209458 b with its parent star providing the source of gravity. As only one side of the planet constantly faces the star, only one side is heated. This leads to huge jets of gas ten times faster than hurricane-force winds on Earth flowing from the hot dayside to the cold nightside. The atmosphere is also hot enough for electrons to be ripped away from some atoms. These charged particles combine with the raging winds and the planet's magnetic field to generate an electrical current in the planet's atmosphere and interior. Like the electricity flowing through a bar heater, this causes the planet's interior to heat up. This heat causes the planet to swell, giving it its inflated radius.[4]

The best models for the inflated radius of HD 209458 b suggest that violent winds are present in its atmosphere. What if we could see more effects of these winds? Could we find evidence of them moving huge torrents of gas around in a planet's atmosphere?

3

A TEMPESTUOUS WORLD

The Muslim month of Ramadan begins with the first appearance of the new crescent Moon. Around 1.7 billion people observe a month-long period of fasting, charity and prayer that begins with the appearance of a sliver of lunar light. They are not alone in having their calendar defined by our celestial neighbour. Hindu, Jewish and Chinese calendars follow the phases of the Moon. These phases also determine the Christian date of Easter.

It is not just the Moon that has phases. When astronomers first looked at Mercury and Venus through telescopes they saw planets that appeared to change shape as they orbited the Sun. Of course, these bodies are not morphing over the course of their orbits. It is just that almost all the light we see from rocky planets in our solar system comes from reflected sunlight. Sometimes we on the Earth are in a position to see more of the illuminated dayside, sometimes we see more of the dark nightside.

In the previous chapter we met HD 209458 b, a planet that gets in the way of the light from its parent star. This causes the star to dim when seen from the Earth once per planet orbit. We simplified things a bit and treated HD 209458 b as almost a perfect black disc that blocks out part of the star's light. Such planets are, of course, real 3D worlds. To show that, let us meet a very similar object.

HD 189733 b was discovered using the radial velocity method, the effect it has on the dark lines of the spectrum of its parent star.[1] It was also found to transit its parent star, blocking out part of its host's light. Its star is a bit cooler than the Sun, but because HD 189733 b is about 14 per cent more massive than Jupiter and in a tight 2.2-day orbit, the planet is still very much a hot Jupiter.

In 2007 astronomers used the Spitzer Space Telescope to observe HD 189733 b's parent star.[2] Sure enough, they saw the transit, the planet getting in the way and blocking the star's light. This transit caused a 2 per cent drop in the brightness of the star. The astronomers kept observing once the transit was over and they saw something strange happen. After the transit the star returned to its normal brightness. Then it started to get even brighter. This was not a huge effect, a change of a little under a quarter of 1 per cent of the star's brightness. But then the star suddenly dimmed again, this time by three-tenths of 1 per cent. This drop was far less significant than the transit and, what's more, it was at the wrong time, almost exactly half an orbit after the transit.

What could have been causing this pattern?

In our solar system we see Venus change in brightness as it moves around the Sun. This difference can be seen with the naked eye and doesn't need a telescope to resolve what phase (fully illuminated, quarter or new) Venus has at any particular point. There are two things that affect this change. First, the further away from us Venus is during its orbit, the fainter it is. Second, the more we see of Venus's dayside, the brighter it is. HD 189733 b is so far away from us that the change in brightness due to it being slightly closer to us during different parts of its orbit can be ignored. However, the brightness of HD 189733 b will change if we see more of its dayside.

When HD 189733 b transits in front of its star, its dayside is entirely facing towards its host. That means the nightside is facing towards Earth as HD 189733 b is located in-between its parent star and us. As the planet moves in its orbit we see more and more of the dayside revealed. This explains the gradual increase in brightness we see. It is not the host star getting brighter; it is that more of the dayside of HD 189733 b is being revealed.

Recall the pattern astronomers saw, a 2 per cent dip in the star's brightness as HD 189733 b transited, a gradual increase in brightness of 0.25 per cent and then a drop in brightness of 0.3 per cent. What could be causing the drop? The clue is in the timing, roughly half an orbit after HD 189733 b transited. At this point the planet will be directly behind its star. Astronomers were seeing HD 189733 b being eclipsed by its parent star.

This seems a complete explanation for what the astronomers saw with the Spitzer Space Telescope. But a few subtle features in the data hint at other bits of information about HD 189733 b.

From just after the point where HD 189733 b transited to just before the eclipse, the increase in brightness of the planet and star combined was 0.25 per cent. Here we went from seeing the star plus the nightside of the planet to seeing the star plus the dayside. So, the difference in brightness between the dayside and nightside of HD 189733 b is 0.25 per cent times the brightness of its parent star. When HD 189733 b was eclipsed by its star the brightness of the star and planet together dropped by 0.3 per cent. At this point neither the dayside nor the nightside could be seen. Combine these two facts and you come to the conclusion that the nightside of HD 189733 b has a brightness of 0.05 per cent times the brightness of its parent star: so, the nightside is not totally dark.

So far we have talked a lot about light coming from stars. Mostly it has been referencing visible light, be it in a spectrum of 51 Peg b or light from HD 209458 being blocked by the planet around it once per orbit. Stars, however, give out a much wider range of light than our eyes can see.

There is a common, and largely untrue, myth that indigenous peoples in the high Arctic have a plentiful array of words for snow in each of their languages. However in English we really do have a tremendous amount of words for the same thing, light. Take a glance at astronomy articles in the press and you will see references to gamma rays, ultraviolet, microwaves, radio waves, X-rays, thermal radiation and, of course, visible light itself. These are all just different names for different types of light. They range in wavelength, which defines the colour of light, from gamma rays to X-rays to ultraviolet to visible light to thermal radiation to microwaves to radio waves.

Across the universe a huge range of astronomical objects give off these different kinds of light in different ratios and through different physical processes. The sky is sending us signals in glorious Technicolor, far beyond what our eyes can comprehend.

The Spitzer Space Telescope is one observatory that helps us see some of the light a conventional camera on a ground-based telescope might miss. It observes infrared light, what you might call thermal radiation. Visible light has wavelengths of 0.4 millionths of a metre for blue light and 0.7 millionths of a metre for red light. Infrared light is redder than this. Telescopes on the ground can see some infrared light but for wavelengths greater than 3 millionths of a metre it is best to use a space telescope. This is partly because the Earth glows in infrared light and partly because the atmosphere blocks lots of light at such wavelengths.

The observations Spitzer took of HD 189733 b and its host star were of light with a wavelength of about 8 millionths of a metre. This was chosen for a reason, one that TV weather presenters have been confusing you about.

In a baking summer heatwave the weather map is covered with reds and oranges. Then, six months later, when the temperature plunges, the same map is covered by blue. This is a familiar picture for people living in temperate countries – it is also completely upside-down from an astronomer's point of view.

Hot stars glow brightest in blue light, cooler stars glow most in red light. The Sun with a temperature of 5,500°C (9,930°F) glows most in greenish-yellow light. A planet like HD 189733 b might be heated up by its star so that it has a temperature of 1,000°C (1,830°F) or so. This means it will glow mostly in the infrared, making it easier to see at wavelengths around 8 millionths of a metre. It's not just the colour of a star or planet that is determined by its temperature, its brightness is too. A hotter body will emit more radiation at all wavelengths than a colder body of the same size. The other factor that determines brightness is size, bigger stars give out more light than smaller ones of the same temperature.

Astronomers have estimates of how much light HD 189733 b gives off both on its dayside and its nightside, and because it transits they also know how big it is. This means they can work out the temperature. They determined that the dayside is about 950°C (1,740°F) and the nightside 700°C (1,290°F).[3] In the infrared the amount of light the planet gives off from its own glow is much greater than the reflected light from its parent star.

Like HD 209458 b, HD 189733 b is so close to its star that it is tidally locked, meaning one side stays facing the star. This is because the star's gravity first stretches the planet into an oblong shape and then drags on the bulge of the oblong, forcing it to face it all the way through its orbit.

This means that the star is always heating the dayside, with the point of the planet closest to the star always experiencing blistering midday heat. But the star is not heating the nightside, so how could it be 700°C there?

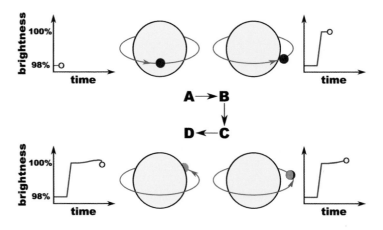

A) HD 189733 b transits its star blocking its light. B) The planet's nightside and the star are seen. C) The bright dayside of the planet is revealed, increasing the observed brightness. D) The planet is eclipsed behind the star.

UP TO THIS point we have come across lots of ways astronomers have looked at planets around other stars. But there is a different way to examine these worlds, by making theoretical models of a planet on a computer.

Astronomers made models of the atmospheres of tidally locked hot Jupiters. They found that the star heating the dayside of the planet but not the nightside drove huge waves and eddies in the planet's atmosphere.[4] These generated and channelled a strong eastward jet of wind around the planet. This jet would move the hottest point in the planet's atmosphere to slightly east of the point where it is always midday. The brightness of HD 189733 b as it orbits its star is at its maximum not right at the point before the eclipse, but one-twentieth of an orbit before. This is the point when we on Earth get the best view of the eastward hotspot.

Now we have met a planet with an atmosphere that redistributes heat by winds. But there are more ways we can probe the atmospheres of these worlds.

4

A GLIMMER OF ATMOSPHERE

You probably know the drill. Perhaps it is a documentary about space, perhaps the introduction to a David Attenborough TV programme. The camera begins above the Earth viewed from space, unlit. As the music swells, the Sun peaks through the Earth's atmosphere illuminating it with a rainbow of colour. Then seconds later a portion of the globe is bathed in sunlight. I am not bringing this up simply to inspire you with the majesty and preciousness of our world. This sequence includes a brief glimpse at an important way we can learn about worlds around other stars.

When we met HD 209458 b in Chapter Two we saw the light from its star was blocked by the planet moving in front of it once per orbit. The star was essentially a glowing sphere and the planet a black circle that gets in the way. The image in this chapter taken from the International Space Station (ISS) provides a good indication of why that model is not entirely right. Yes, the Earth is a black disc obstructing the Sun's light, but it also has a thin ring of atmosphere around it that sunlight can pass through.

This chapter's planet, WASP-19b, is another hot Jupiter, this time in a nineteen-hour orbit around a yellow-orange star a bit cooler than the Sun.[1] The planet itself is about 11 per cent more massive than Jupiter and is, like other hot Jupiters, puffed up with a radius 40 per cent bigger than the dominant world in our solar system.[2]

One thing that is different about this planet is how it was found. The previous three planets were found by the radial velocity method, in which the shift in the dark lines in a star's spectrum shows the star is being moved around in an orbit by a planet. WASP-19b was not found by lots of painstaking spectral measurements of individual bright stars. Instead astronomers took images covering huge chunks of the sky using small telescopes each looking at a different bit of the heavens. When I say small telescopes, I mean small. The WASP array that found this planet was a collection of high-quality telephoto lenses, some purchased from eBay.[3] These lenses had wide fields of view and allowed astronomers to measure the brightness of lots of stars over big areas of the sky. When a planet in orbit round one of these stars transited, its star's brightness would dip. The detector behind the telephoto lens would notice this fading of the star and astronomers would monitor it further to see if the planet transited again one orbit later. This was how WASP-19b was found.

Let us go back to the image of the Earth with the Sun poking through our planet's atmosphere. Say you were far away from the Earth, far enough away so that the Earth looked much smaller than the Sun. Now say from this position that you see the Earth transit the Sun, like we see transits from planets around other stars. The rocky Earth would block out the light from the bit of the Sun it was obscuring, but a thin ring of atmosphere around it would have the Sun shining through.

WASP-19b is a gas giant planet, not a rocky world like Earth. However, it still has a part that is dense and opaque enough to block out all of its star's light during a transit surrounded by a thin ring of atmosphere illuminated by the star's light.

What will an atmosphere do to starlight passing through it? First, remember how the atoms in the star's atmosphere took

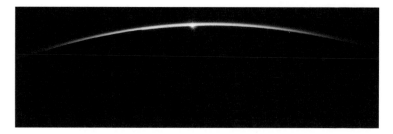

Seen from the International Space Station (iss), the Sun rises through the thin sliver of atmosphere above the Earth.

nibbles out of the star's light before it escaped into space: the atoms in the planet's atmosphere do the same thing. Not just atoms, molecules too. Molecules are the gluttons of spectroscopy, filling their faces with photons and taking much wider bites from the spectrum than atoms do.

You can see a few other things atmospheres do to starlight by looking out of the window during daylight. Why is the sky blue? Because the atoms and molecules in the Earth's atmosphere scatter blue light from the Sun. Look at a blue patch of sky. Light from the Sun has hit that bit of atmosphere and scattered in all directions. You see it as blue because some of the light was scattered towards your eyes and because the atoms and molecules scatter more blue light than red. On a polluted day you might see haze or smog. This is caused by aerosol particles suspended in the air scattering sunlight. This smog also scatters light, deflecting more blue light than red light. In a planet's atmosphere these hazes will scatter blue starlight into space. This means that during a transit less blue starlight passes through the planet's atmosphere to reach an observer on Earth.

Clouds are another example of scattering, deflecting light from the Sun that passes through them. These typically block light irrespective of its colour. A layer of clouds blocks out starlight and cloaks the atmosphere below, making it hard to observe anything beneath the clouds.

All three of these things can happen in an exoplanet atmosphere and all three will do the same thing, reduce the amount of starlight that comes directly in a straight line from the star, through the planet's atmosphere to the observer. This means that while the middle bit of the planet will block all of the star's light when it transits, the thin ring of atmosphere will only block and scatter some.

This means that the atmosphere of the planet affects the depth of the little dip in the star's brightness we see when the planet transits. The starlight we receive at the telescope during this transit is encoded with information from the planet's atmosphere.

WHAT WOULD astronomers need to do to collect and interpret the information hidden in the transit of WASP-19b? They need to measure the transit depth (the amount of light blocked out by WASP-19b) in many different colours of light. This could be by taking a spectrum of WASP-19b's star a couple of times a minute or measuring the star's brightness through many different-coloured filters. Recall that the colour of light depends on its wavelength.

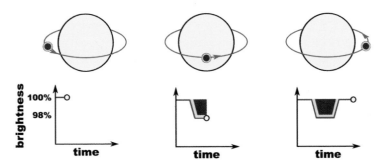

One part of WASP-19b (shown here in dark grey) blocks its star's light completely. Starlight passes through the ring of atmosphere around it (shown in orange). This ring of atmosphere also blocks light from the star, but it blocks more at some wavelengths than at others.

All of these techniques are looking for a transit depth that varies with wavelength.

Like the Earth, there will be a bit of WASP-19b that blocks out all of the light from its parent star. This means there is a minimum depth for the transit. On top of this, the thin ring of atmosphere around the planet will absorb and scatter starlight. The amount of this absorption and scattering will depend on the wavelength of the starlight. This means that at some wavelengths more light will be absorbed and scattered, and thus the transit will be deeper than at other wavelengths.

Astronomers didn't pick WASP-19b as an observing target at random in the hope of finding something. They have help from theorists. These are astronomers who build mathematical and computer models of different types of astronomical objects. The observers then take these models and use them both to interpret their results and to plan new observations.

The models theorists build inform astronomers' choice of observing targets. A narrow ring of planetary atmosphere will not change the depth of a transit very much. A big puffy plane-tary atmosphere, on the other hand, would catch more starlight and affect the transit depth much more. There is also the consid-eration of how bright the planet's parent star is. Astronomers study stars by measuring how much light comes from them. The more light they get from a star, the more accurate the brightness measurement. In the case of a planet around a faint star, astron-omers would need to observe for longer during a transit to get enough light for an accurate measurement. This would mean they would be able to make fewer measurements during a transit of a planet around a faint star than during the transit of a planet round a bright star. Having more measurements helps with another problem: stability.

HUMANITY HAS AN incredible ability to think of a variety of different names for the same thing. The name of a bread roll changes across Britain from buttery to roll to cob to bap to barm cake. There is also a popular children's game that changes name from country to country: be it 'telephone' in the USA, 'quiet post' in Germany or 'broken telephone' in Malaysia, the game relies on the same thing, changing words in a sentence as it passes from person to person.

Astronomy has a similar process when it comes to light from a star. The light enters the Earth's atmosphere, where some of it can be absorbed or scattered. Then it enters the telescope and travels through a series of lenses and mirrors before finally striking the detector. At each one of these stages light can be altered, much like the sentence in the popular children's game. And how the light is altered can subtly change over time. The atmosphere, the telescope, the instrument and the detector are all go-betweens passing on the message, and sometimes they can be shifting and unreliable.

Astronomers have an advantage. Unlike the kids in the playground who only have one sentence to go on, astronomers can observe more than one star at a time. These stars should have intrinsic brightnesses that are independent of each other. So, if their brightnesses change at the same time, then it is one of the unreliable go-betweens passing on the message that is causing the variation. This change can then be corrected, leaving only the changes in the target's signal itself.

Armed with models of hot Jupiter atmospheres, several groups of astronomers have measured the transit depth of WASP-19b across different wavelengths. They also monitored the brightness of reference stars to check how our atmosphere, the telescope, the instrument or the detector were affecting the measurements of WASP-19b's parent star and corrected for any effect they saw.

What the teams saw was that when WASP-19b transited across its star it blocked out 1.9 per cent of the light at all wavelengths. This was due to the bit of the planet that blocks out all the light from the star. However, on top of that 1.9 per cent minimum transit depth there were some wavelengths at which more light was absorbed. These differences were not huge, the thin ring of atmosphere around the planet blocked out less than a quarter of 1 per cent of the star's light. But there were definite variations at wavelengths where molecules of water ate away at the star's light.[4] One team also found evidence for titanium oxide in the planet's atmosphere and a slope with more and more blue light being scattered at shorter, bluer wavelengths.[5] Another team, however, with a more detailed model of the light coming from WASP-19b's parent star, recently found neither a blue slope nor titanium oxide.[6]

What does this glimpse at the WASP-19b's atmosphere tell us? We can see that there is water vapour in the atmosphere, which means that it is both cool enough for water to form and the right temperature for it to take big bites out of the star's spectrum. Having the spectral signatures of water does not mean that this planet is a world with oceans like the Earth. Giant planets like Jupiter have water vapour in their atmosphere and there are even cool stars that show water features in their spectra caused by steam in their atmosphere. The recent finding of a flat spectrum at shorter, bluer wavelengths suggests that WASP-19b is covered with a blanket of high-altitude clouds.

We have met four thoroughly weird worlds so far, all of them similar to Jupiter in mass and all in very close orbits to their star. This shows how alien and unlike our solar system planets around other stars can be. But we have not dealt with one question, how did these massive planets get to be in such short orbits around their stars?

5

A WORLD IN REVERSE

Perfection is a trifle dull. It is not the least of life's ironies that this,
which we all aim at, is better not quite achieved.
W. Somerset Maugham[1]

Perfection doesn't exist.
Andrés Iniesta[2]

So far in this book we have been dealing with stars and the
planets in orbit around them. We have been treating the star
as a glowing, perfect orb with a few references to the star's atmosphere. The idea of the perfection of celestial bodies was popular
in Western astronomy until the seventeenth century. Heavenly
bodies were viewed as, well, heavenly and thus perfect.

Luckily that is not true. The first observations of the Sun
and Moon using a telescope showed one body marked with dark
evolving spots and another scarred by craters and jagged mountains. It is fortunate that such celestial perfection does not exist,
because it would be a trifle dull. Stars are far, far more interesting than a perfect glowing sphere. And it is stars' imperfections
that can help us learn about how hot Jupiters became in their
current, weird orbits.

IF THE Sun is not some perfect celestial orb then it must follow
the laws of physics. This produces a couple of problems.

First, the Sun shines. This means it is radiating heat and thus must either be constantly cooling down or have some sort of energy source. Second, the Sun is a massive ball of material. Something must be keeping it from collapsing under its own gravity.

These two problems produced headaches for nineteenth-century scientists. They knew that the Earth was old from the rocks around them scarred by glaciers and volcanic eruptions. But they were mostly familiar with fuels like coal and found that these were not a realistic source for the Sun's heat, as they would not produce enough energy for the Sun to shine brightly for hundreds of millions of years. The Scottish physicist Lord Kelvin, a scientific titan of the age, spent decades lurching between theories in a fruitless quest to explain the Sun's source of power. At one point he even theorized that a constant and fantastically intense shower of meteors powered the Sun.[3]

It was not Kelvin's fault that his theories about the Sun turned out to be wide of the mark. It took until the twentieth century for two important pieces of the puzzle to appear that helped solve the mystery of the Sun's power. First, Cecilia Payne-Gaposchkin showed that the Sun is mostly hydrogen and helium, not rocky like the Earth, as some had previously thought. Second, nuclear fusion was discovered. These two things combined to give us a basic model of how the Sun and other stars work.

In the middle of a star we find a nuclear furnace. This is heated to more than 10 million degrees Celsius (18 million °F) by the crushing mass of the star's material pushing down on it from above. The core is so hot that hydrogen nuclei can fuse together. This produces the energy to power the star and to stop it from collapsing in on itself.

Each star burning hydrogen in its core balances the energy produced by the nuclear reactions with the energy radiated away into space. This energy also heats the core and supports it against collapse under its own gravity.

The more massive the star, the more material pushing down on the core, the hotter the core, the more reactions, the more energy the star radiates away. These more massive stars also have hotter outer layers. Hotter stars are blue, colder stars are red, like the reverse of a TV weather temperature map, as mentioned.

Energy is produced in the core of stars like the Sun. This energy is then radiated away from the outer layers of the star. You might have noticed the gap in the logic here: how does the energy get to those outer layers?

There are two ways this energy can be transported. One is radiatively, by which, roughly speaking, each layer heats the layer above it (that is, the layer further from the core). The other way to transport energy is convection. Just as when hot air rises from the tarmac on a baking hot day, hot material from deep inside the star rises and transports heat with it.

Stars can be a mix of radiative and convective zones. Massive stars (with masses over 1.5 times the mass of the Sun) have convective cores and radiative envelopes. Envelope here means the part of the star extending from the core to just under the 'surface' of the star. Not that the Sun or any other star has a hard crust or liquid surface. The nearest the Sun has to a surface is the part of the star above which the material is thin enough for light to escape into space. The Sun and stars of similar mass have radiative cores and convective envelopes. The lowest mass stars, less than half the Sun's mass, are entirely convective.

Remember how the first telescope observations of the Sun were spotty and imperfect? This spottiness showed something else: by tracking the spots as they gradually move across the face of the Sun, we see that the Sun rotates. The axis of rotation

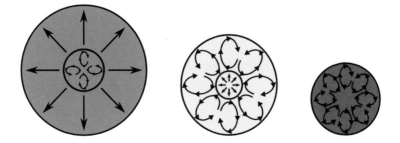

The structures of a high-mass (left), Sun-like (middle) and low-mass (right) star. The straight lines show radiative zones, the curved convective. Diagrams not to scale.

is roughly aligned with the orbital axes of the planets. This means that the Sun is spinning round in the same direction as the planets are orbiting it.

This rotation also has another consequence: half of the Sun is moving toward us as it rotates, half is moving away. Remember how 51 Peg b's star became bluer as it moved towards us and redder as it moved away? The same applies to different parts of the Sun. The part of the Sun that moves towards us gets a little bit bluer, the part that moves away gets a bit redder. For each dark line in the Sun's spectrum, each nibble taken by the atoms in its atmosphere, part of the light we see will be bluer coming from the bit of the Sun moving towards us, part will be redder coming from the bit of the Sun moving away from us. This makes each line in the spectrum wider and not a narrow, sharp line as you might expect. The faster a star rotates, the broader the lines in its spectrum.

This chapter's planet, HAT-P-7b, is a hot Jupiter, and a big one at that, with a radius that is 40 per cent bigger than Jupiter and a mass 75 per cent greater. The planet orbits a star about 1,000 light years away that is a little hotter than the Sun. Like all other hot Jupiters, this planet is in a tight orbit around its star, going around it every 2.2 days.[4]

HAT-P-7b was discovered by the transit method because it blocks out some of the light from its primary once per orbit. If astronomers had a ridiculously powerful telescope and could resolve HAT-P-7b's star, they would see a circular disc akin to how the Sun appears from Earth. As the planet passes between an observer on Earth and the star, it first covers the edge of the disc, then moves across, finally reaching the other edge. It might go through the middle of the disc, it might just graze the edge.

HAT-P-7b will cover different bits of its star at different times in the transit. If the star is rotating then one part will be moving towards us and one part will be moving away from us. If during its transit HAT-P-7b blocks light from the bit of the star moving towards us, then it will remove some of the blue-shifted light from all the star's spectral lines. As the lines would now have more red-shifted than blue-shifted light, it would look like the star had started to move away from us. Conversely, if HAT-P-7b blocks out the part of the star moving away from us, then red-shifted light will be removed from each of the star's spectral lines. Hence each line would have more blue-shifted light and it would look like the star was moving towards us.

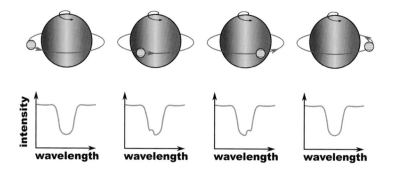

A rotating star is transited by its planet. Here the star's spin is aligned with the planet's orbit. The transit covers different parts of the star at different times, changing the shape of each spectral line.

Most planets orbit their star in the same direction as the star spins. This is true for planets in the solar system. For such a planet you would first see a shift to redder wavelengths during a transit, then to bluer wavelengths. For a planet that orbits perpendicular to the direction of its star's spin the effect would be very different. It might only cross parts of the star that are red-shifted or blue-shifted, so you would only see a movement in the spectral lines in one direction.

HAT-P-7b does something very odd indeed. First it covers a bit of the star that is moving away from us, then a bit that is moving towards us. HAT-P-7b appears to be orbiting the wrong way around its star!

How does a planet end up orbiting the wrong way around its star? Theoretical models of planet formation predict that it's hard to form a big planet close to a star. HAT-P-7b was probably formed further away from its star and orbiting in the same direction as its star rotated. Then some other body orbiting HAT-P-7b's star started pushing the planet around. It may have been another planet orbiting at larger distances from the star, it may have been a distant binary star companion to the star, but something started giving HAT-P-7b gravitational kicks. Over time these kicks stretched HAT-P-7b's orbit so that it went from circular to elliptical, then shrunk it so HAT-P-7b moved closer to its star. Finally the orbit was twisted so that HAT-P-7b was not orbiting in exactly the same direction as its star was spinning. Eventually this twisting would become so great that the orbit was perpendicular to the direction of the star's spin. Then after even longer the orbit would be flipped over so that HAT-P-7b was going the wrong way around its star.

Actually HAT-P-7b has not completely flipped its orbital direction. It is flipped by 120 degrees, so not quite the full 180-degree flip, but still enough for it to orbit in reverse.[5]

It is worth noting that, as is often the case in science, there are competing explanations for the orbits of hot Jupiters like HAT-P-7b. The most common alternative theory is that the planet migrated while it was still in the disc of material it formed from around its parent star. Again another planet or star in the system could be responsible for distorting HAT-P-7b's orbit, either by twisting the disc it formed from or by warping HAT-P-7b's orbit after it formed.

SINCE THE late 1990s astronomers have managed to measure alignment between a planet's orbit and its star's spin axis for more than two hundred planets. These results show an interesting thing: hot Jupiters around cold stars generally have orbits aligned with their star's spin, while those around stars hotter than 5,800°C (10,472°F) mostly are not.[6] One of the best theories astronomers have to explain this is that planets that have been pushed into misaligned or reverse orbits then start turning their stars around.[7]

On Earth we are used to seeing the tides, which are caused by the Moon's gravity stretching the water in the Earth's oceans so that it collects in two bulges, one facing towards the Moon, the other on the opposite side of the Earth. A similar thing happens for hot Jupiters: they slightly stretch their parent stars so a small bulge is pointing towards them.

This bulge follows the planet around as it orbits. It also interacts with the rest of the star, dragging on it. This process leads to the planet losing some of its orbital energy and moving closer to the star, but also dragging the star so that it begins to rotate the same way as the planet orbits.

Planets orbiting in reverse can turn their stars around so they spin in the same direction as the planets' orbit. But why do we only see planets like HAT-P-7b orbiting the wrong way around hot stars?

The answer comes back to the frothing convective envelope that cool stars have and hot stars do not. The churning envelope generates a strong magnetic field. Each star sends out a wind of charged particles. The Sun's wind is the source of the spectacular aurorae near the Earth's poles. The strong magnetic fields generated by a cool star grabs onto this wind of particles. As the wind moves away from the star it remains locked to the magnetic field. Like a spinning skater throwing out their arms to slow down, the wind carries away rotational momentum from the cool star and causes it to slow its rotation.

It is much easier for a hot Jupiter orbiting in reverse to turn around a slowly spinning cool star than a fast-spinning hot star. Hence a hot Jupiter in a reverse or misaligned orbit round a cool star can turn its star around to spin in the same direction as its orbit, while a similar planet orbiting a hot star would not.

SO FAR, we have met five very alien worlds and learned about their atmospheres and orbits. But are there worlds more like the Earth out there?

TOWARDS EARTH

6

A FLASH FROM DARKNESS

'Don't play with your food': a common phrase used by parents across the world. Unfortunately, some of the students at one of the top U.S. universities hadn't been listening. Around the year 1920, students at Yale University started playing games with left-over food containers. They would toss the pie tins from the local bakery, the Frisbie Pie Company, around campus. Eventually a California company that had been making flying discs under a different name decided to market their products as 'Frisbees', possibly in reference to the game the Yale students played.[1]

History is littered with examples of items designed for one purpose being used for something different. The next planet we are going to meet was found using a technique that was developed to find a completely different type of astronomical object.

Astronomy is one of the more media-friendly sciences. Gorgeous vistas of huge clouds of gas and shimmering stars are a great way to attract the attention of a reader flicking through a newspaper on the way to work or browsing a news site during a break. Three topics in astronomy stand out as ones that attract a lot of press attention: planets, black holes and dark matter. The story of this chapter's planet takes in all three.

The last of these three subjects is one of the main puzzles of modern astrophysics. Astronomers can measure the mass of a galaxy or of a cluster of galaxies by looking at the motions of the stars and gas in them. In spiral galaxies like the Milky Way we can see a disc of stars rotating round the centre of each galaxy. By measuring the dark lines in the spectra of these galaxies we can work out how fast different parts of that disc are moving. The faster the stars in the disc of the Galaxy are orbiting, the more mass is pulling them round in their orbits. Astronomers can also map out the distribution of stars in each galaxy and estimate the total mass of all those stars, too.

Astronomers can compare the orbital speed of the stars in each galaxy with the orbital speed they would expect to measure if the stars were being pulled round solely by the combined mass of all the stars in that galaxy. It turns out that the stars in galaxies are orbiting faster than expected. This indicates that there is more mass pulling them around in their orbits. This missing mass is commonly referred to as dark matter. Soon after the discovery of dark matter, astronomers started to come up with ideas for what it could be. One theory was that there was some undiscovered particle that had mass (and hence gravitational pull) but did not interact with the atoms and light in the universe in other ways. Another was that galaxies were full of dense, compact objects that emitted no light or very little light. These could be old, cold white dwarfs (the remains of long-dead stars) or they could be black holes, objects from which light cannot escape.

So how do you study something that does not emit light? Well, you make use of the one thing you know it has, a gravitational pull.

IN 1919 two teams of astronomers travelled to remote locations to observe a solar eclipse. One team headed to the island of Príncipe in the Atlantic while another went to Sobral in northern Brazil. Neither team was there to simply gaze in awe as the Sun disappeared behind the Moon. They were looking for something else, stars.[2]

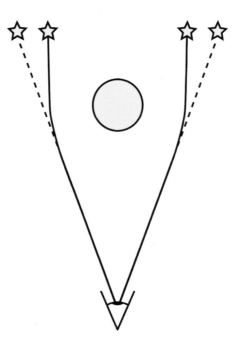

Light from stars is bent by the Sun's gravity (solid lines). This makes it look to the observer on Earth (here shown by an eye) like the stars have shifted in position. These altered positions can be viewed during a solar eclipse.

During a total solar eclipse, the sky is dark enough to see stars. 'So what?' you might say, 'it's dark at night too, just look at the stars then.' The reason looking at stars during a solar eclipse is interesting is that astronomers can perceive stars that appear to be close to the Sun in the sky. The light from these stars has to pass near to the Sun before reaching Earth. The path that light takes is slightly bent by the Sun's gravity. This alters the stars' positions in the sky. At other times of year the Sun is in a different part of the sky so astronomers can measure the true positions of the stars at night. Then during an eclipse, they can measure how much the positions of the stars have been altered by the distorting effect of the Sun's gravity.

Both expeditions measured the distortions of the positions of stars close to the Sun in the sky. Both found that the bending of the stars' light by the Sun was in line with what Albert Einstein had predicted in his General Theory of Relativity. This spectacular result blanketed news-stands worldwide and Einstein moved from being a lauded academic to a global celebrity. All because the Sun acted like a giant magnifying glass.

It is not just the Sun that has a gravitational magnifying glass: all stars do, and so do all dead stars and every black hole. These magnifying glasses come in handy when looking for dark matter in our Galaxy.

LOOKING UP at the night sky you would be forgiven for thinking that the stars are fixed in place. Quite the opposite is true. The Milky Way is like a city at rush hour, with each star hurrying around the Galaxy on its own orbit. The bulk pattern is that the stars in the Galaxy mostly orbit in the same direction. But, like the hustle and bustle of traffic in the city, some stars are cutting across the paths of other stars, while some are overtaking and some falling behind. The speeds involved are mind-boggling. The stars close to the Sun on this great galactic go-around are moving relative to the general flow of stellar traffic at speeds ten times faster than a fighter jet. And that's on top of a flow of traffic that is ten times quicker than those speeds. The reason the stars look stationary is that they are so far away that it would take centuries to move enough on the sky for you to perceive it with your naked eye.

Say you were looking across many lanes of traffic at a car in the furthest lane from you. Every so often another car would get in the way and block your line of sight to the far-away car. Going back to our interstellar traffic, the chances of a star passing in front of another star and blocking that star's light are very small. But as we've seen, stars can do something else: bend light like a giant magnifying glass.

So how does all this relate to dark matter or, more to the point, exoplanets? One of the explanations for the missing matter in the universe was that it was made up of dense objects like black holes that do not give off light. Those black holes will be moving round the Galaxy just like stars do. Now say one of those black holes passes in front of a star astronomers are observing. The black hole, like a star, is a gravitational magnifying glass. This means its gravity will focus the light from the star of which it is passing. While the effect is not exactly the same as focusing the Sun's light with a magnifying glass to start a fire, the outcome is similar. The star being lensed appears to get brighter.

Astronomers looking to find out if dark, massive objects like black holes were the main constituents of dark matter started doing surveys of the sky. They would stare with their telescopes at a dense field of stars in the Large Magellanic Cloud, a galaxy very close to the Milky Way. They hoped that every so often they would see the brightening of one of those stars caused by a black hole in our Galaxy passing in front of the star. This is known as a microlensing event.

The surveys for these microlensing events did not detect enough black holes and other dark, compact objects to explain dark matter. Hence our best guess at the moment is that dark matter is made of some exotic particle that rarely interacts with atoms and light.

Microlensing as a technique might not have shown that black holes make up a significant amount of the dark matter in our Galaxy. But, like the Frisbie pie tins that were tossed around Yale's campus, there was another use for this technique.

Stars passing in front of other stars can cause a microlensing effect. The star that passes between us and the background star is typically referred to as the lens. Planets also have mass, so can act as lenses, although their lower masses mean the size of the microlensing effect is smaller. However, if a planet is in orbit around the lens star then this acts as a magnifying glass on top of another magnifying glass, temporarily boosting the magnification power and making the background star that is being lensed brighter.

In July 2005 a Polish telescope in Chile was monitoring an area of sky densely packed with stars. They noticed one star was getting brighter and so on 11 July issued an alert: this was a microlensing event. Telescopes from various international collaborations started to monitor the star. These observatories were in Chile, Hawai'i, New Zealand, Tasmania and Western Australia. By spreading the observations around the world, the event could be monitored constantly, even if it was cloudy or daytime at some sites.

The event peaked on 31 July and the brightness of the background star that was being lensed started to decline. Then, just as it looked like the event was winding down to a quiet end, something extraordinary happened. First one telescope in Chile saw an increase in the star's brightness, and then another observatory in Chile saw it, then New Zealand, then Western Australia. This was what they had been looking for, a small boost in the magnification caused by a planet orbiting the lensing star.[3]

Let us meet the world that caused that boost in brightness late in the microlensing event. It is called OGLE-2005-390L b and it is not like the other worlds we have come across so far. This planet was not the first found by microlensing: two Jupiter-like worlds were discovered using this technique shortly before OGLE-2005-390L b was found. OGLE-2005-390L b is a lot less massive than Jupiter; it is even less massive than Neptune. From the characteristics of the microlensing event, astronomers estimated that OGLE-2005-390L b has a mass of 5.4 times the mass of the Earth, less than a third of the mass of Neptune.

So again, we have a world for which we have no analogue in the solar system. It falls between the terrestrial planets (like Earth and Venus) and the ice giants (like Uranus and Neptune). Worlds like these could either be described as super-Earths or mini-Neptunes.

OGLE-2005-390L b is, like the two microlensing worlds found before it, in a very different orbit than the hot Jupiters found in the last section. The magnification boost that powers the micro-lensing technique is most significant for planets in orbits a few Earth–Sun distances from their star. This orbital distance compares to 51 Peg b, which orbits its star at one-twentieth of an Earth–Sun distance. The characteristics of OGLE-2005-390L b's microlensing event show that it orbits its parent star at a distance of at least 2.1 Earth–Sun distances.

The hot Jupiters we met in the last section were all in orbit around stars similar to the Sun. OGLE-2005-390L b is in orbit around a small star that has a mass less than one-quarter of the Sun's mass. This star is hence much cooler than the Sun. These cool red stars are far more common in the Galaxy than stars like the Sun.

The microlensing technique for planet detection follows a single event that lasts for a few weeks with a short burst of magni-fication caused by the planet that lasts a few days. Once the event is over it is not possible with currently available technology to observe the planet again. OGLE-2005-390L b appeared to us in a flash that lasted for a day, but that burst of brightness is all we will see from this world. There will be no follow-up observations of this planet to determine its radius or to search for its atmosphere.

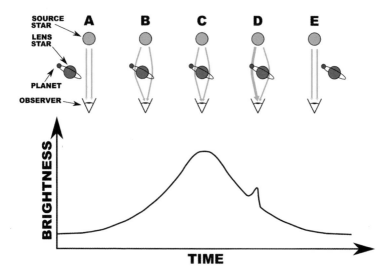

The change in brightness of the background source star for microlensing event like OGLE-2005-390L b. A) The microlensing event starts; B) the source star is lensed and magnified, increasing in brightness; C) the star's brightness peaks; D) as the brightness is declining the planet provides an additional microlensing boost, briefly increasing the brightness; E) the microlensing event ends.

That does not make planets discovered by microlensing useless. This technique provides a window into the population of small planets in moderately wide orbits, objects other techniques struggle to identify. Microlensing finds the planets other techniques struggle to reach. We know there are other worlds around other stars, but how many planets are there in the Galaxy? What type of orbits are they on? Are worlds like the Earth common or rare?

So how can we find more small worlds, and work out how many worlds like Earth there are? First, we will need to meet a very special spacecraft, one that will eventually give us the answer.

7

A WORLD YOU CAN'T SET YOUR WATCH BY

You probably know that person, the one in your social circle who is habitually late for everything, forever rushing through the door looking flustered. Or perhaps you know someone who, regular as clockwork, will turn up fifteen minutes early when you aren't quite ready yet. The planet we are about to meet in this chapter has timing that is all over the place. Sometimes it is far too early, sometimes far too late.

So far, we have met planets that were discovered through three different methods: radial velocity, transits and microlensing. These planets were discovered by astronomers, many of whom started their careers in fields of astronomy that used smaller, often older telescopes than the state-of-the-art large observatories that dominated other fields of astronomy.

Just over a decade after the discovery of 51 Peg b, specialist planet-hunting spacecraft began to appear. First, the French-led COROT in 2006, followed in 2009 by NASA's Kepler mission.

The Kepler spacecraft sat in space and used an on-board telescope to constantly point at one patch of sky. This allowed it to monitor continually the brightnesses of 150,000 stars. If a planet transited in front of one of these stars then Kepler saw a dip in that star's brightness. Even though Kepler was operating in space, the stars and planets it observed are so far away that it essentially saw the transits from an almost identical perspective to a telescope on Earth.

A few chapters ago we came across the idea that signals from stars are passed on by unreliable intermediaries, the atmosphere, the telescope, the detector. As a spacecraft Kepler does not have to deal with the Earth's atmosphere interfering with its measurements of stars' brightnesses.

Kepler's serene staring in space is in contrast to the chaotic environment faced by the astronomer the mission was named after. Johannes Kepler accurately calculated the orbits of the planets around the Sun. He had to take a break from his research to defend his mother, Katharina, from charges of witchcraft.[1] He also had a colourful patron, the eccentric Holy Roman Emperor Rudolf II, a man obsessed with astrology and alchemy who was famously painted by Arcimboldo as a collage of vegetables. Rudolf collected fantastical art and oddities and even had a tiger roaming his palace.[2] The latter curiosity is thankfully one barrier modern astronomers do not have to face in the quest for funding for their research.

The Kepler mission got off to a steady start. First it observed a number of previously known hot Jupiters. Then it discovered a few hot Jupiters of its own. The observations were promising, showing that the telescope was able to measure the size of the transit dip for these planets very accurately.

The ninth planetary system Kepler identified, known as Kepler-9, was the first it saw that contained multiple planets. Systems with multiple planets had been found before by radial velocity searches, but this was the first discovered by Kepler.[3]

This system consists of two gas giant planets, with masses between a third and half the mass of Saturn. One planet, the subject of this chapter, has an orbit with a period a little over nineteen days, the other gas giant has an orbital period a little under 39 days. The system also has a smaller transit signal from what looks like a super-Earth in a close orbit. This planet wasn't properly characterized in the initial discovery paper so we will leave it to one side and focus on its two larger siblings.

The transiting planets we have met so far have a clockwork regularity to them. Once per orbit, on time, the planet would pass in front of its star and the dimming of the star would be observed from Earth.

Something was strange about Kepler-9. Unlike in previously observed systems, the two planets were not transiting at the predictable, clockwork times you would expect. Sometimes the inner planet (Kepler-9b) could be twenty minutes late transiting and the outer planet (Kepler-9c) could be an hour early. Then, a few months later, the reverse situation would be seen. Kepler-9b would be twenty minutes early in its transit time and Kepler-9c would be an hour late. What was causing this variation, this transgression from the typical transit tick-tock timing?

Setting problem sheets for physics students can sometimes be tricky. Have I made sure that they have covered the material needed for these questions? Will all the students understand the set-up of the problem? (I remember a German student in my undergraduate year being confused by an exam question involving a cricket ball being bowled and answering the question as if a ten-pin bowling ball was being bowled.) Finally, is the problem solvable?

A generally quick question might involve a planet orbiting a star. I might ask students to work out the orbital period of the planet, or maybe to estimate the radial velocity variation the planet would induce on the star. Things get a lot harder, though, when you add another planet to the system. This takes the problem away from being a simple five- or ten-mark question into something that's beyond the scope of a quick spot test of student knowledge and ability.

When there are two or more planets in orbit around a star there is no elegant orbital solution that can be quickly calculated with pen and paper. The system is a mess of planets orbiting the star and the planets also exerting tiny gravitational tugs on each other. However, among the chaos, nature can arrange itself into a configuration that is in its own way graceful.

The two giant planets in the Kepler-9 system both orbit their star. The inner planet orbits faster so will sometimes overtake the outer planet on its orbit. It is at this point that the two planets are closest; hence this is the moment where each feels the other's gravity the most. At this point the planets give each other gravitational kicks, subtly changing each other's orbits.

In some systems this kind of interaction can lead to kicks adding up, one after the other, with the orbits of the planets being drastically altered. Perhaps eventually one planet will get so many kicks that it will be thrown out of the planetary system altogether. Some planets may also find themselves in orbits where they receive small kicks from other planets but where these kicks eventually average out over longer cycles of several orbits. There are also special orbits that planets in a system can end up in. These are known as orbital resonances, with the two planets ending up in orbits where the period of one planet is in a simple mathematical ratio with the period of the other planet. For instance, in our solar system Pluto (a dwarf planet) orbits the Sun with a period that is $3/2$ times the period of Neptune's orbit.

The Kepler-9 system with its two giant planets in relatively close orbits around their star has evolved, through billions of years of gravitational kicks, into a system where Kepler-9c has roughly twice the orbital period of Kepler-9b. The gravitational interactions between the two sometimes cause the planets to wander slightly from these special orbits, sometimes with one going faster around their star, the other slower, sometimes with the situation reversed.

It is the subtle changes in the planet's orbits induced by the gravitational kicks that cause the transit times to deviate from the expected regular pattern. This is not a simple astronomical curiosity, this is a useful technique for characterizing planets.

The size of the gravitational kicks in the Kepler-9 system depends on the mass of each planet. This means that astronomers can use computer models of the orbits of the two planets, along with the measurements of the transit timing variations to constrain the planets' masses.

The team that discovered Kepler-9 used these transit timing variations to determine that both planets had masses of around half the mass of Saturn. They were also able to measure the change in the radial velocity of the parent star in the system. These measurements confirmed that the planet masses determined by transit timing variations were accurate.

The star in the Kepler-9 system is bright enough for accurate radial velocity measurements. Unfortunately this is not true for all planet-hosting stars. To measure a radial velocity astronomers split the light up into a spectrum. This is in effect lots of data bins of different colours. The brighter the star, the more light each bin in the spectrum has and the more accurate the measurement in that bin will be.

Imagine surveying the ages of sixteen people and splitting them into eight age ranges. You might get two per age range on average, but might by chance also get four aged forty to fifty and zero aged thirty to forty. How confident would you be in saying that nobody in the population you were sampling was thirty to forty years old? Do the same survey for 1,600 people and you would typically get two hundred people per age range. Here you might be more confident in the significance of having four hundred forty to fifty year olds and zero thirty to forty year olds. The same is true for bright stars; if there is more light in each spectral bin, astronomers can have more confidence in the measurements in those bins.

Even when observed for long periods of time using the largest telescopes, many faint stars do not have enough light in each spectral bin to measure their radial velocity accurately. Hence the masses of planets around them cannot be determined by looking at the changes in the radial velocities of these stars. Transit timing variations are crucial for measuring the masses of such planets.

As well as giving us the mass of the planets, transit timing variation tells us something else about the Kepler-9 system: that it is real. Transit timing variation is something astronomers would expect to see in real planetary systems. The fact that these variations are observable in the Kepler-9 system gives us evidence that Kepler-9 is a real planetary system. Unfortunately, that is not always the case. The universe is full of weird and wonderful objects and many of those can mimic the signal from a transiting planet.[4] So how can astronomers tell if a star that shows periodic dimming has a planetary system around it and how can they be sure the universe is not just tricking them into believing that?

Let us go back to what a transit signal actually is, the drop in a star's observed brightness. It is measured as a fraction of the star's light, light blocked divided by the total amount of light an observer would receive from the star when no planet was transiting. But what if the full area of the transiting object was not blocking light from the star? It might be that a much larger object goes in front of the star but only a small part of it actually gets in the way and blocks the star's light on its way to Earth. This grazing transit would mimic a transit by a planet by blocking the same amount of starlight that is heading towards Earth.

There is also the possibility that what Kepler is seeing is not light from one star but light from several stars close together in the sky. These stars could be hundreds of light years apart but appear close enough in the sky to fit into one Kepler pixel. In this case their light would be measured as if it was coming from one star.

Let us examine a particularly pathological case that could mimic a transiting planet. We have a star that is observed by Kepler that is five hundred light years away from the Earth. Behind that star and 4,500 light years further away there is a binary star system, two stars in orbit around each other. If each of the components of the binary star system on their own were observed from Earth each would have a brightness fifty times smaller than the foreground star. However, by chance the star Kepler is observing and the binary star fall in the same Kepler pixel. This means their

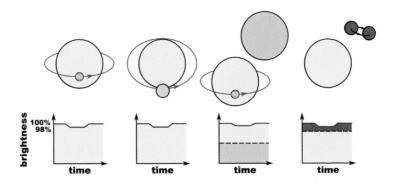

False positives and spurious measurements for transit surveys like Kepler. Far left, a real planet: a planet transits a star causing a transit depth of 2 per cent. Centre left, a grazing transit: a larger body transits a star but only a small portion of the body covers up the star. This leads to a transit depth of 2 per cent. Centre right, a blended binary: a planet transits a star but a second star is observed in the same Kepler pixel. This makes the transit depth look smaller and leads to an incorrect estimate of planet radius. Far right, background eclipsing binary: the light from a bright foreground star dilutes an eclipse in a background binary system. This leads to what looks like a 2 per cent transit depth.

brightnesses are measured together, as if they are coming from one star with each component of the binary star system contributing 2 per cent of the light. Let us say that at one point during the binary star system's orbit one star passes between the other component of the binary system and Earth. This means one star in the binary system would eclipse the other, blocking its light from reaching Kepler. As each star in the binary system contributes 2 per cent of the total brightness in the Kepler pixel, Kepler would see the total brightness in that pixel drop by 2 per cent. This would be repeated each time the binary star system had an eclipse. A periodic drop of 2 per cent in the brightness of a Sun-like star is exactly the sort of signal Kepler would receive from a giant planet in orbit around that star.

Having more than one star in a Kepler pixel can also give astronomers the wrong answer on the size of a planet. Say there are two stars of equal brightness in one pixel. If one of them has a hot Jupiter planet in orbit around it, its brightness will periodically drop by 2 per cent. But the total drop in the combined brightness of the two stars would only be 1 per cent. This is the drop Kepler would measure. Astronomers would get the light-blocking area of the planet wrong by a factor of two and underestimate its radius by about 30 per cent.

The same issue can also arise if there is only one star in each pixel but that star is much larger in physical size than astronomers thought. An Earth-like planet transiting the Sun might cover 0.01 per cent of its areas and block 0.01 per cent of its light. A Jupiter-like planet orbiting a giant star ten times the radius of the Sun would cover 0.01 per cent of its area and block 0.01 per cent of its light.

Astronomers need to do extensive follow-up observations to confirm and characterize individual planets. Measurements of the shift in the parent star's radial velocity can determine a planet's mass. They can also rule out a grazing transit by a larger body as this body would have a higher mass. Taking high-resolution images of the parent star can sometimes spot other stars nearby that could have fallen in the same Kepler pixel and distorted the transit depth. Modelling of the transit light curve in different colours of light can also rule out the possibility of an eclipsing binary.

With all this in mind, let us meet a possibly rocky planet and its somewhat different sibling.

8

A CONTRASTING SIBLING

We all know siblings who are very different from one another. They could be celebrities or people in our everyday lives. It could be Erik Nielsen, the former Deputy Prime Minister of Canada, and his brother the deadpan comic actor Leslie Nielsen. It could also be the quiet bookish kid and his ruffian sporty sister, or the school's head girl and her rakish layabout brother.

Planets are often different from their siblings too. Our own solar system has a diverse range of planets: rocky, gassy and icy. Sometimes sibling planets can appear similar, but on closer inspection be very, very different.

This chapter's world, Kepler-36b, is a transiting planet discovered by the Kepler mission in 2012 with a 13.8-day orbit around its star. Interestingly it has a quite different sibling, Kepler-36c, which has a 16.2-day orbit.[1]

After Kepler first identified these planets, astronomers set out to characterize both worlds as well as their parent star. For most planet host stars this would mean estimating the star's temperature and brightness, and using a model to estimate its radius. This radius, combined with the size of the dip caused by the planet's transit, would give the astronomers a measurement of the planet's radius. For the study of the star in the Kepler-36 system the astronomers were able to get a more accurate measurement of stellar radius by exploiting an interesting physical effect. This star rings like a bell.

Watching the news, you might see a report of a disastrous earthquake that has sent buildings tumbling and left thousands homeless. It is strange to think that even if you lived on the other side of the world the nearest seismology station to you probably detected the earthquake. The sudden shift in the Earth's crust sends sound waves pulsing through our planet's molten interior that can be detected on the other side of the world.

Stars don't have crusts but sound waves can still travel through them. Like a musical instrument, there are certain harmonic frequencies where a star will 'ring', a bit like a bell. These special frequencies change with the size and internal structure of the star. Harmonic pulsations in a star can cause its brightness to vary. This means that measuring the brightness of a star over time and then looking at the pattern of how the star brightens and dims can tell astronomers how the star is pulsating and ultimately allow them to measure its properties.

Using Kepler's long-term measurements of the variations of Kepler-36's brightness, astronomers were able to determine that it is a bloated subgiant star that is about as hot as the Sun, but has a radius 1.6 times bigger. This radius estimate, combined with the measured transit depth of both planets in the system, gives radii of 1.48 Earth radii for Kepler-36b and 3.68 Earth radii for Kepler-36c.

These two planets orbit in an orbital resonance. Kepler-36b goes around seven times in its orbit in the same time Kepler-36c goes around six times. The two planets give each other subtle gravitational kicks, which move them in and out of exact resonance, shifting the times each transits the star. This means that the planets can have their masses measured by transit timing variations. Kepler-36b has a mass of 4.3 Earth masses and Kepler-36c has a mass of 7.7 Earth masses. These masses, when combined with the measured radii, result in something remarkable. The two planetary siblings, in such similar orbits around the same star, have densities that differ by a factor of eight.

IN THE last two chapters we have met super-Earths. Both OGLE-2005-390L b and the possible third planet in the Kepler-9 system fall into this classification. When we talk about hot Jupiters it is easy to think of the solar system planet that would look similar: Jupiter. But a super-Earth is by definition not the same as Earth, it's much bigger. The next biggest planets in our solar system after Earth are the ice giants Uranus and Neptune. The latter of these is 3.9 times the radius and seventeen times the mass of Earth. Neptune is physically very different from Earth. The Earth is rocky and has a thin atmosphere primarily composed of nitrogen and oxygen, but while Neptune also has a rocky core, this lies beneath a layer of superionic water and a thick atmosphere of hydrogen and helium. The two planets in the Kepler-36 system are bigger than Earth but smaller than Neptune. Are they bigger versions of the Earth, mini-Neptunes or something else entirely?

To answer this question, astronomers made a series of computer models of each planet. Each model had several components that went together to make each planet. For example, a model could consist of a planet with an iron core surrounded by a mantle of rocky material. Different materials have different densities: iron is denser than rock, which is denser than water, which is denser than an atmosphere of hydrogen and helium. This means that two planets with the same mass can have very different sizes if they are made of different materials. A planet that is a solid ball of iron would be much smaller than a planet of the same mass that is made up entirely of water.

The astronomers allowed the fractions of the planet's mass made up of each component to vary and then estimated the radius each model planet would have. This radius could then be compared with the observed radii of the planets in the Kepler-36 system. Does a 99 per cent iron and 1 per cent rock model give a radius for Kepler-36b that matches the observations? No? Then how about a 98 per cent iron and 2 per cent rock model? No? Well then, next model.

Kepler-36b is a bit denser than the Earth. The models of Kepler-36b's interior showed that if it were made of rock and iron then it would have an iron core making up 30 per cent of its mass with the other 70 per cent being rocky material. The relative fractions of these components are roughly similar to those found in the Earth.

There is also another possibility, that Kepler-36b is a water world, a rock world covered by a substantial layer of water. Running a model with three components (iron core, rocky mantle, water outer layer) astronomers found that Kepler-36b has at most 23 per cent of its mass in a thick layer of water. The best match to the observed radius was found with a model with a watery envelope making up 13 per cent of the planet's mass.[2]

None of the eight major planets in our solar system is a water world. While the Earth has oceans, these make up a very small amount of our planet's mass. The solar system does give us some water worlds to study. Several moons of Jupiter and Saturn have a rocky core and a thick layer of water. However, these moons are in cold parts of the solar system so have icy surfaces. Their interiors are heated by tidal stretching caused by their planet's gravity. This leads to a subsurface ocean, something space missions to Jupiter and Saturn have found evidence for in several moons such as Europa, Ganymede and Enceladus. However, Kepler-36b is hot, about 800°C (1,472°F). This means that much of its water could be in the form of a steamy atmosphere.

Astronomers also tested if Neptune-like models would fit Kepler-36b. There the model planet was made of a rock/iron core, a layer of water and then an atmosphere of hydrogen and helium. In this case the models showed that at most 1 per cent of the planet could be an atmosphere of lighter elements. This is much less than the fraction of Neptune's mass that is made up by hydrogen and helium.

So, Kepler-36b is not a mini-Neptune, but either a rock and iron terrestrial planet or a water world, something not seen among the major planets in the solar system.

The same modelling technique came up with some very different answers for Kepler-36c's composition. This planet has a much larger radius than its sibling Kepler-36b. Kepler-36c is so puffy that it must have an atmosphere made up of lighter elements such as hydrogen and helium. Beyond that it is hard to determine what is beneath the atmosphere. Kepler-36c could have rocky material under its atmosphere or it could be a rocky core surrounded by water and an atmosphere on top.

Both interior models for Kepler-36c need to reproduce the observed radius of the planet. Rock is very dense, so a rock-only interior model would need a thick hydrogen and helium atmosphere (about 9 per cent of the planet's mass[3]) on top of the core to match the observed radius. By contrast, an interior made of rock and water would be less dense, meaning perhaps a per cent or two of the planet's mass would need to be made up of a hydrogen and helium atmosphere.

Here we have two worlds: the one that is the subject of this chapter is either a rocky world or a water world. Hence it is truly a super-Earth. Its more puffy, bloated sibling has an atmosphere of hydrogen and helium and is likely a mini-Neptune. How did two such contrasting siblings form?

It is possible that the two planets formed in very different places around their parent star or that they formed at different times early in the star's life. This could lead to a difference in composition. There's also another possibility, however, that they both formed with similar compositions but that those compositions changed with time.[4]

Both planets in the Kepler-36 system are close to their star and move in short orbits. Models of planet formation suggest that bigger planets form further away from their parent star. In Chapter Ten we will meet a still-forming planet and will examine the formation of planets in a bit more detail. In this chapter, however, we only need to know the basic fact that planets form in discs of material around young stars. Once planets are formed they can change their orbits. This could be because they receive subtle gravitational kicks from other bodies in the system (like HAT-P-7b did) or because they interact with material left over from the planet-forming disc. This means planets can initially form far away from their parent star and then ultimately move into closer orbits.

The planets in the Kepler-36 system could have formed further away from their star and then moved into shorter-period orbits. The disc from which they are formed is made of gas (like hydrogen and helium) and dust (tiny specks of rocky or metallic material). This dust can stick together to form pebbles, which bind to form planetary embryos.

Far away from the parent star, the disc is cool enough for any water or carbon monoxide to freeze and turn to ices. This material can stick to the still-forming planets. Gases such as hydrogen and helium can also be pulled in by the forming planet's gravity.

An artist's impression of the mini-Neptune Kepler-36c viewed from the possibly rocky Kepler-36b. At its closest approach Kepler-36b would be two-and-a-half times the size of the full Moon viewed from Earth.

Once this process of formation and migration was complete around Kepler-36, it is possible that the two planets in the system looked quite similar: big rocky cores, maybe some water around that core and then a thick, puffy atmosphere of hydrogen and helium. There would have been one crucial difference; Kepler-36c would have been far more massive than Kepler-36b. This is important, because planets are fighting a constant battle against their stars.

Stars might give planets heat and light, but they can also play a destructive role. As well as this warmth and visible light, stars send out high-energy electromagnetic radiation like X-rays and a stream of fast-moving particles. The closer a planet is to its star, the more of this destructive bombardment it faces. Over time this can strip away parts of a planet's atmosphere.

Models of the Kepler-36 system show that even if Kepler-36b was formed with a thick hydrogen and helium atmosphere, this would have been stripped away over the course of a billion years. Astronomers estimate the Kepler-36 system is about 7 billion years old, so there's been plenty of time for this atmospheric erosion.

Kepler-36c would have faced a similar blast of X-rays and high-energy particles from its star. However, Kepler-36c started out with a higher mass. This meant that it had a stronger gravitational pull so could better resist the brutal ransacking radiation from its parent star. Even after nearly 7 billion years of bombardment it still retains its envelope of hydrogen and helium, making it a puffy mini-Neptune.

The Kepler-36 system may also demonstrate something important about the population of exoplanets in general. Looking through a sample of 2,000 planets with orbital periods under one hundred days found by the Kepler mission, astronomers found lots of puffy mini-Neptunes like Kepler-36c and lots of super-Earths like Kepler-36b. Between these two populations there was a big gap, with few planets with radii about 1.8 times that of Earth.[5] This may be because worlds less massive than Kepler-36c simply cannot hold on to their lighter gases like hydrogen and helium when faced with blasts of X-rays from their parent stars.

Here we have met a large, probably rocky planet that, in contrast to its sibling, is likely to be a larger version of the Earth. But what of planets like the Earth? Do they exist around other stars? And how many of them are there?

9

A WORLD LIKE OURS?

One of the most frequent clichés in journalism is declaring a person or group at the start of their career to be the next incarnation of some legendary figure. Chameleonic young actresses are trumpeted as the next Meryl Streep, tricky diminutive footballers are cast as the next Lionel Messi and charismatic up-and-coming politicians are declared to be the next Barack Obama. The same affliction has affected the descriptions of music bands. Since Beatlemania a whole slew of groups, from the Bay City Rollers to Oasis to One Direction, have been hailed as the next Beatles. Astronomy suffers from a similar affliction.

The discovery of a new planet often appears in the media as the 'most Earth-like planet ever discovered' or even under a banner headline posing the question 'is this the next Earth?' This is understandable; one of the prime motivators of the search for other planets is to find out how common Earth-like worlds are. The problem is that 'Earth-like' is a very flexible term. Just as there are a whole range of factors on which newcomers can be compared to the Beatles, Streep, Messi or Obama there are many different ways a planet around another star can be compared to the Earth. A planet could be similar to the Earth in size, composition, atmospheric chemistry or in surface temperature. With the huge number of factors that make our world unique, however, it is highly unlikely that we will ever find a planet exactly like our own.

That does not mean it is pointless to talk about 'Earth-like' planets. It is just important to understand that there will be no exact twin of Earth. We can, however, learn how many planets are similar to the Earth in size and composition and in the amount of radiation each planet receives from its parent star. In the future we will also be able to characterize the atmospheres of these planets. Some planets similar to Earth may be able to host life. There will

" HE SAYS HE'S FROM AN EARTHLIKE PLANET. "

This won't happen. But we can still find planets that are similar to the Earth in size, composition and even temperature.

be a whole range of diverse rocky worlds. Searching among this plethora of peculiar planets for one that looks just like ours seems like British tourists on holiday searching for a restaurant that sells fish and chips, instead of sampling from the wonderful variety of global cuisine.

This chapter's planet is Kepler-10b, a world discovered in 2011, one year before the Kepler-36 system.[1] Kepler-10b is an 'Earth-like' planet, but in this case the comparison is literally rock solid.

The Kepler spacecraft used the transit method to find Kepler-10b, a planet in a twenty-hour orbit around a star a little cooler and less massive than the Sun. The Kepler-10 system is about five hundred to six hundred light years away from the Sun. Kepler-10b is small, with a radius 1.47 times that of Earth.[2] Measurements of the shift it induces on the dark lines in its parent star's spectrum show that the planet has a mass of 3.3 times the mass of Earth. This mass and radius give a density for Kepler-10b that puts it very firmly in the terrestrial planet category. It is made of rock and iron, just like Mercury, Venus, Earth and Mars.

Kepler-10b was the first exoplanet found to be unambiguously rocky. One other planet found before it, COROT-7b, was subsequently confirmed to be a terrestrial planet, but only after the discovery of Kepler-10b.

Kepler-10b is Earth-like in its composition and orbits a Sun-like star. It is, however, very different in another respect, its surface temperature. Kepler-10b is so close to its star that it is tidally locked, with one side permanently facing its parent body, and receives a far higher amount of radiation than the Earth does. These two factors lead to it having a temperature on its dayside of about 1,500°C (2,732°F), meaning that it cannot host life that is based on liquid water. One could either consider this to be a hot Earth-like planet or a super-Mercury.

The discovery of Kepler-10b was a significant moment, the first rocky planet found around another star. Soon Kepler identified other planets similar to Kepler-10b. These discoveries would help astronomers to finally answer a question that had been driving the exoplanet field since it began: how many rocky worlds similar to Earth are there?

Pick up a newspaper or turn on the television to watch the news and you will likely be confronted at some point by the results of an opinion poll. These tell you which politician's career is on the up, which policies are popular with the public or even which crisp flavour has recently been crowned the nation's favourite. In making these polls survey companies will try their best to canvass a representative sample of men and women, people of different ethnicities, ages and political beliefs.

In astronomy it would be great to do such a survey. First find a representative sample of a thousand stars that contains just the right proportion of hot stars and just the right proportion of cool stars. Then find all the planets around each star and use that number to work out the fraction of stars that have planets similar to the Earth in size and temperature.

The problem with this approach is that the techniques astronomers use to detect planets around other stars do not detect all the planets around every star they look at. Take the radial velocity method. Remember how the size of the shift in the dark lines in a star's spectrum depended on the mass of the planet orbiting that star and the angle the system was viewed at. A planet in a face-on system (a system that looks like a fried egg stuck to a wall with the star in the yolk and the planet going around the white) would show no shift in the dark lines of the star's spectrum. Hence astronomers using the radial velocity method would look at such a system and detect no planets.

An artist's impression of Kepler-10b, a rocky world blasted by the intense heat of its parent star.

A similar principle applies to transiting planets viewed by Kepler. We can only see transits from a planet that has an orbit that crosses our line of sight to the planet's parent star. If the planet never comes between its star and us we will never be able to detect it with the transit method.

There are also more subtle biases in the transit method. As the diagram opposite shows, closer-in planets are more likely to transit their star than further away ones. This means we are more likely to observe closer-in planets with the transit method.

Planet detection and characterization with the transit method requires multiple transits to succeed. Kepler's main observing run staring at a single field of stars lasted four years. In order to measure the orbital period of a transiting planet at least two transits must normally be observed. For Kepler to see a planet transit its star twice during its observing run, that planet must normally have an orbital period of four years or less. A planet in a five-year orbit would transit, at most, once in this time, as would a planet on a ten-year orbit.

Multiple transits also help to improve the detectability of a planet. Say you were tossing a coin and got heads. You would not be that shocked, as the probability of getting heads is the same as getting tails. Now say you landed ten heads in a row, you would start to question if the coin you were tossing was fair. The same happens with the transit method. Observations of the brightnesses of stars are noisy. Even if a star has a steady brightness, the brightness values astronomers will measure will bounce around that constant level. An observation of a star might show a measurement that was just below the normal value for that star. That might just be down to noise on the measurements, making it a coin toss if one particular measurement is above or below the average value. But what if, regular as clockwork, the brightness of a star dipped slightly once every six days for four years. That pattern would be hard to explain by chance. Astronomers can use algorithms to spot patterns like this in Kepler data where statistically insignificant individual transits can add up to a statistically

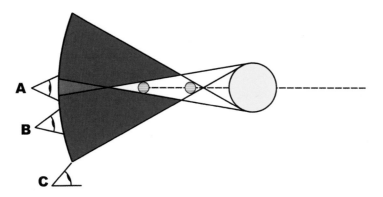

A system with two planets is observed from three different positions. The top observer (A) can see the full disc of both planets transiting their star. The middle observer (B) can see only the inner planet transit. The bottom observer (C) sees no transits.

significant pattern. The more transits Kepler observes, the easier it is to spot such patterns.

With all this in mind, astronomers set out to measure the number of stars that have Earth-sized planets in Earth-like orbits around them. Armed with the same Kepler data set, different teams of astronomers focused on the Earth-sized planet occurrence rate around various types of stars. Some looked at the number of Earth-sized planets around stars like the Sun, others looked at the number of Earth-sized planets around cooler, smaller, redder stars called red dwarfs. Red dwarfs are the most common type of star in the Galaxy. In the part of the Galaxy surrounding the Sun red dwarfs outnumber yellow dwarfs like the Sun fifteen to one.[3] Hence the number of Earth-sized planets around such small red stars will have a big effect on the total number of Earth-sized planets in the Galaxy.

As well as trying to pin down the number of Earth-sized planets, the teams looked at how many of those planets are in the habitable zone around their parent stars. Roughly speaking, a planet is in the habitable zone if the heat and light it receives from its parent star give it a surface temperature that could sustain the existence of liquid water. As mentioned, being able to host liquid water gives a planet the potential to host life.

There is one minor problem with estimating the number of Earth-sized planets in the habitable zone: Kepler did not find any. It found some slightly larger rocky planets in the habitable zone and it found some very uninhabitable Earth-sized worlds. This meant that the astronomers needed to make some assumptions about the distribution of planet radii and the distribution of the planet orbits around stars.

Differing assumptions is one reason that two teams came to different values for the number of Earth-sized planets in the habitable zone around stars similar to the Sun. One team found that about 22 per cent of Sun-like stars will have Earth-sized planets in the habitable zone,[4] another team found 2 per cent.[5] Both of these numbers shoulder fairly large uncertainties. A team looking at Earth-sized planets in the habitable zone around red dwarfs found that about 16 per cent of these stars host a rocky world that could have liquid water.[6]

These measurements tell us how many Earth-sized planets there are in the habitable zones around other stars. They do not, however, tell us if those planets have life. The question of intelligent life on other planets is often written down as the Drake Equation, which breaks the occurrence rate of other technologically advanced life in the Galaxy down into several questions, starting with how many planets are there, how many could host life and how many of those go on to host life? Kepler has provided an answer to questions about how many Earth-sized planets there are in the habitable zone. However, it will be much harder to determine if such worlds can or in fact do host life. To go a step further, determining if these worlds host technologically advanced life will be even more difficult.

The statistical analysis of Kepler data shows that planets similar to Earth are not found around every star, nor are they exceedingly rare. The Kepler planet data set also allows astronomers to investigate which stars are more likely to host planets and to delve into the statistics of planetary systems.

It appears that there may be some order to the family portraits of exoplanet systems. If two planets orbit a star, those planets are likely to be of fairly similar masses.[7] This also holds for systems with more than two planets, with the second planet from a star tending to be similar to the first, the third being similar to the second and so on. We see this in our solar system, with Venus being rocky like Mercury, the Earth rocky like Venus and so on out to the pair of ice giants, Neptune and Uranus. There are breaks such as Jupiter being very different to Mars, but the solar system planets are not randomly ordered, nor are worlds in other planetary systems.

From such studies we also know that planets are more common around stars that contain more heavy elements like iron and silicon.[8] This makes sense; planets start out as seeds of rocky material made from heavy elements and grow by accreting more heavy elements. Stars that form with more heavy elements will also contain more heavy elements, the building blocks of planets, in the discs around them. This will make it easier to form planets around these stars.

Astronomers also know, thanks to Kepler, that stars in some binary systems (two stars orbiting each other) are less likely to host planets. It has been suggested that this is because the gravitational pull of a stellar companion could disrupt and suppress the planet-forming process in a disc around one or both of the stars in the binary system.[9]

So, what is the planet-formation process? How do planets form? And can we see them forming?

BIRTH

10

AN UNSEEN EMBRYO

Deep in a swirling mass of material around a young star, something is lurking. It cannot be seen, but its existence can be deduced from what it has fed on, what it has devoured. This is not a beast from a low-budget horror film complete with unconvincing rubber creatures and buckets of fake blood. This is a young, still-forming planet.

Secondary school teaching is often a thankless task. It is not easy to balance the pressures of constrained budgets, ambitious parents and the need to get students past a battery of tests and exams. On top of all that there is the difficult matter of getting pupils engaged in what can sometimes be dry material. Classical mechanics is one such area, a myriad of balls colliding on friction-less snooker tables and projectiles being fired at arbitrary angles that just happen to give round numerical answers.

One rare beacon of interest in this otherwise dull field involves a member of the class sitting on a revolving office chair while holding large textbooks at arm's length. When the spinning student pulls the weighty tomes towards their body, their rotation speeds up rapidly. This principle also applies to spinning figure skaters and children moving inwards on a revolving roundabout in a playground. For students wishing to get out of the classroom, the latter makes a great outing suggestion to an easily led physics teacher. The former is only a viable class trip suggestion for a particularly gullible teacher.

A star begins its life like a student with outstretched arms and a hefty textbook in each hand. Well, it is several billion kilometres across and rather than being an enormous teenager, it is a clump made of gas and the sort of sandy and metallic particles that astrophysicists refer to as dust. The clump of gas is itself the result of a long process where a huge cloud of gas, tens to hundreds of light years across and with as much material as tens of thousands of stars, collapses. Rather than an orderly, spherical collapse, the frothing, turbulent motions within this huge mass of gas drive it into a series of dense, knotty filaments. Lying inside these cosmic tangles are clumps like the one we will be following, many of them on the way to forming stars.

After such a violent process our clump is not sitting perfectly still, it is spinning very slightly. Gravity starts to do its work and the clump begins to collapse. Like our spinning student pulling their textbooks towards their body, the clump begins to rotate more rapidly. Material begins to concentrate in the centre of the clump, the seed for a forming star. Further out, the spinning of the clump speeds up to the point where some of the material does not fall into the centre but instead forms a rotating, pancake-like disc around the forming star. This disc is made of the same material the clump started out with: gas and fine specks of sandy and metallic matter.

In the centre of the disc, the densest part of the now-collapsed clump becomes hot enough to switch on as a star. Further out, the disc of material does not shine in visible light like a star, but glows in a different way.

The dusty, sandy material in the disc around our young star has some residual heat from its formation and gets some heat from the star in its centre. Even so, it can still be as cold as $-250°C$ ($-418°F$). Such a bitter bulk of material still emits light far beyond what our eyes can see, in a region of the electromagnetic spectrum called the submillimetre. This brings us to an image of this chapter's planet, or rather an image of this planet's home.

The image shows the area around HL Tau, a star that sparked into life less than a million years ago.[1] It lies in the Taurus star-forming region, one of a number of huge complexes of gas and forming stars within a thousand light years of the Sun. What you see in the image opposite is not a picture taken with a conventional camera, it is instead taken by a series of antennae that resemble satellite dishes. More than fifty of these are spread over an area a kilometre or so across, forming one large observing facility, the Atacama Large Millimeter Array (ALMA).

Each ALMA dish is equipped with a receiver that detects the submillimetre radiation that is reflected towards it. Every dish will detect subtly different signals from an astronomical source, some dishes receiving a particular signal slightly earlier than others. These small differences are then used by a powerful computer at the heart of ALMA that combines the signals from all the dishes to reconstruct an image of the source being observed.

The result of this complex process linking high-tech pieces of equipment spread across the Chilean desert is a series of concentric circles around HL Tau. The whole disc is roughly two hundred Earth–Sun distances across with seven dark rings separated by about ten Earth–Sun distances each. These dark rings represent gaps in the disc where the material has either disappeared or been pushed away, but what has it been pushed away by?

To tell the story of these gaps in a disc billions of kilometres across, we need to move down to the scale of the tiny grains of dusty material that inhabit it. These minuscule morsels less than one-tenth of a millimetre across are blown and buffeted around the disc by the gas surrounding them. The howling winds in the

The dusty disc around the young star HL Tau. The gaps indicate areas of the disc that appear to have been cleared out by embryonic planets.

disc as the gas whips around the central star combined with turbulent flows and eddies force the dust into traps, areas of the disc where the dust begins to concentrate. Here the concentration of dust is large enough that the grains start to collide regularly. Sometimes these grains stick together, sometimes they are smashed into smaller grains. Over time these processes lead to some grains becoming larger.

THE BASKING shark is one of nature's odder-looking creatures. With a mouth that is 90 centimetres (3 ft) wide, it does not mercilessly bite down on its prey like some of its more iconic, dread-inspiring cousins. Instead it slowly drifts through the water with its gaping mouth gobbling up small fish and sea creatures. The larger and wider the mouth a basking shark has, the more food it is capable of guzzling.

A planetary seed in a disc feeds in a similar way to a basking shark. Perhaps the size of a speck of dust, it moves through the disc and sometimes meets other dust grains. If one of these other grains collides with it, the two may stick together. As it grows larger the area of the tunnel it bores through the disc material gets wider and draws in more and more matter.

Our gluttonous planetary embryo has now grown from its initial size of less than one-tenth of a millimetre, past the metre or so scale of a basking shark's mouth to be a few kilometres across. Now its gravity is strong enough to pull in the dust grains whizzing about the disc towards its surface. This further increases the radius of its jaws as it swims along.

Other planetary embryos have been growing at the same time, reaching a range of different sizes. So sometimes our forming planet devours a rain of pebbles, sometimes it absorbs larger bodies tens of metres across. But it is not just dusty material that is being accreted. Simple compounds such as water and carbon monoxide will be so hot close-in to a young star that they are in gas form. Further away from the parent star these will be in the form of ice. This gives planets in the outer reaches of planetary systems more solid material to gobble up.

The diet of a planetary embryo is not confined to solid material. Despite the punishing heat and blasting winds of material being given off by the parent star, the disc our planetary embryo inhabits has still managed to hang on to a large amount of gas left over from its formation.

Gas is not the stickiest of things. When a chunk of solid material impacts our now-forming planet, it will crash into the possibly molten surface, being easily absorbed by it. Gas on the other hand doesn't come slamming down in a big blob. Each atom in the gas is buzzing around with its own velocity, some moving with a low speed relative to the planet, some moving much faster. If an atom of gas falls into the gravitational jaws of a planet, it might be moving fast enough that it will whizz past without being slowed down and pulled into its primordial atmosphere. A bigger planet will have a bigger gravitational pull and will be able to slow down and pull in more gas.

Atoms of gas also bump into each other all the time and thus change their velocities. Sometimes these collisions can speed up a particular atom a great deal. The atom may now be moving fast enough to fly away and escape the planet. The bulkier the planet, the higher the speed an atom needs to be kicked to get out of its gravitational pull. This effect means that massive planets are better at holding on to their atmospheres than smaller planets. Atoms and molecules of lighter elements like hydrogen are also more likely to be kicked to very high speeds. This means that planets that are more massive can hang on to lighter elements more easily. As lighter gases like hydrogen and helium are more common in a disc around a young star, more massive planets can accrete more massive atmospheres.

The disc of material around HL Tau has a series of gaps, making it resemble an archery target. Each of these gaps could be the orbit of a forming planet that has gobbled all the gas and dust in its path. Between these gaps, the remaining gas and dust is shepherded by the forming planets' gravities into tight rings.

It may look like an ordered, stable situation has been reached for HL Tau and its disc. This, however, is only a snapshot in time, a portrait of a constantly churning, evolving system. The star in the centre is continually sending out relentless, withering radiation. This rips the gas out of the disc, dragging or evaporating some of the dust with it. The planets themselves do not inhabit the sort of clockwork orrery beloved by Georgian intellectuals. Their orbits are chaotic and slowly changing over time. This brings some of the planets close enough that they collide and coalesce. Others meet a much larger sibling at a high relative velocity and are whipped out of the disc into wider orbits or even ejected into deep, cold, interstellar space.

By the time HL Tau reaches the age of our Sun in around 5 billion years, most of the signs of the disc will be gone. There may be some rubble left over, asteroids that did not amalgamate to form larger planets. Further out there may be a collection of icy bodies resembling the cometary nuclei we believe are floating on the fringes of our own solar system. But in orbit around the host star will be one or more planets. Perhaps some will be gas giants like Jupiter, perhaps some will be smaller.

The HL Tau system shows the dark paths swept out by young, still forming planets. But what do young planets look like? Luckily there is a way to see them directly.

11

A WORLD THROUGH THE BLUR

Perched on the top of a Christmas tree is one of the most recognizable symbolic shapes, a five-pointed star. We know that stars are huge spheres of gas that, when viewed over astronomical distances, look like infinitesimally small dots. But when a star is drawn in art it is very often an object with four, five or six pointy bits sticking out. Mario doesn't gain invincibility by grabbing a glowing sphere. Nobody decorates their Christmas tree with an infinitesimally small dot. Banners fluttering from flagpoles from Samoa to Somalia to Suriname display an array of pointed shapes in every colour. So why are stars so often represented like this?

There's a simple reason why stars are commonly represented as pointed shapes: when you look at a bright star directly with your eyes it looks like it has shards of light sticking out from a central point. This is because your eye is an optical system. Optical systems of different types spread out light from a dot-like point like a star in different ways. In your eye, tiny imperfections in the lens mean that stars have bright spikes when you look at them.

It is the same for telescopes. The pattern a star's light is spread out into is determined by the shape of the main telescope mirror and all the structures used to support the various mirrors that the light bounces off before it reaches the detector. This means that most big telescopes will take pictures where the bright stars have their light spread out, so they look like a blob with four spikes sticking out at right angles to each other. For dimmer stars those spikes are too faint to appear on images but the central blob remains, meaning they look like circles. The size of the circle each star's light is spread out into depends on the size of the telescope the picture of the star was taken with. Larger telescopes can make pictures where the stars look like smaller circles because they do not spread the stars' light out as much.

So bigger and bigger telescopes should make stars look smaller and smaller in images – except that an irritating inconvenience gets in the way, what you are breathing now, the atmosphere.

The images on walls of ancient Egyptian tombs are a mix of the figurative and the realistic, the mythical and the everyday. Alongside the jackal-headed gods and pharaohs smiting enemies a tenth their size are scenes of hunting and fishing. One image in the tomb of Usheret in Thebes, dating from around 1430 BCE, shows a man on a boat spearfishing. This practice goes back into the distant human past before the development of farming or large-scale settlements. Naively one would think spearfishing would be easy; you see a fish in the water, you point your spear at it and you stab – except the fish is not actually where you think it is.

You are probably familiar with a prism or a lens bending light. The material in the lens or prism has a slightly different refractive index (a property relating to how a substance interacts with light) to air. This difference in refractive indices causes the path of a beam of light to bend. For our ancient Egyptian fisherman the same principle applies. Light coming from the fish is bent when that light passes from one material (water) to another (air). This means that the fish is at a subtly different position, slightly below the image the fisherman sees.

The refractive index does not just change between materials such as water and air, it can change within them. Even on a calm day the atmosphere is a seething mass of currents and eddies. Each of these eddies will have a slightly different refractive index so will act as a lens, bending the light by a tiny amount. The cumulative effect of the constantly changing, frothing vortices in the upper atmosphere and the effect they have on light that passes through them leads to the position of a star viewed through a telescope moving around constantly. The changes in position are tiny, typically an arcsecond, which corresponds to the angular size of a pound coin viewed from 4.5 kilometres (2.8 mi.) away. Over the time an astronomical image is taken (from tens of seconds to several hours) this process blurs out the perfect, point-like image of a star. Like a moving sparkler tracing out a pattern in a long photographic exposure, the atmosphere smears the star's light around. Such smearing provides a limit to the resolution of observations. This means that using conventional observing techniques, even with the biggest telescopes at the best observatories with the calmest atmospheres, astronomers struggle to produce images where the stars are blurred out into circles less than half an arcsecond in size.

So will a typical image of a star be a neat, round circle, about an arcsecond across, with well-defined edges? Well, again, not quite. It might look like that to the eye, but the light from the star does not just stop at a hard limit, it keeps going further out. The bigger the distance from the central point of the stellar image, the less and less starlight there is – meaning the light tapers off rather than stopping at a hard edge.

Of course, brighter stars have more light to spread out, so even several arcseconds away from the star's position there is enough starlight to drown out the relatively small amount of light you would get from a planet orbiting that star. That is a problem if you want to find planets. Say an astronomer on a planet around Sirius, the brightest star in the night sky and only 8.6 light years away, tried to observe Jupiter. They would see that it was separated from the Sun by just two arcseconds.

If our atmosphere spreads out the light from stars and we need that light to be spread out less to spot planets around those stars, then the obvious place to go is space. First, though, there is only one Hubble Space Telescope, so it is a lot harder for astronomers to get time to observe with it than with other telescopes. Hubble is also a relatively small telescope; it's not even in the top forty largest optical or infrared facilities. This means the maximum theoretical resolution is just a quarter of what a large ground-based telescope could achieve if atmospheric blurring could somehow be removed. Hubble's smaller size also means it would need to observe for longer to collect as much light as larger, ground-based telescopes.

So how could we remove the effect of the atmosphere for ground-based telescopes? Often the best way to solve a problem is to pick a situation where you know the answer and work backwards from there. You know that a star is not really moving around the sky quickly; it is just the turbulence in the atmosphere causing it to jump skittishly. So you could pick a really bright star with lots of light and use a high-speed camera to constantly take images of the moving image of that star. The positions of the star in these images can then be used to bend a continually moving flexible mirror to correct the position of the star on the telescope detector so that it sits in the same position all the time. This removes most of the atmospheric blurring. The process of detecting the corrections required and implementing them to get to a corrected image is known as adaptive optics.

Once we are left with the sharpest possible image of the star, that's good enough, right? Alas no, the light from the star is still spread out enough to drown out the most close-in planets. To fix this, astronomers need to consider the pattern made by a star in an image.

As mentioned already, the light from a star is spread out on an image, a combination of effects caused by the atmosphere and the telescope. Unfortunately, even after correcting for atmospheric blurring, there is still a lot of spread-out light from the star. Luckily there is a way to get rid of this: work out how a typical star's light is spread out and use this model to remove the light from the star from the image. One way to do this is to pick a star that is both bright and close to the target star on the sky, take an image of that star and use that image to make a model of how a star's light is spread out by the telescope and the atmosphere. Then subtract this modelled pattern from the image of the target star. This removes the nasty, spread-out light around the target and could possibly reveal a planet.

Now that we know how to get observations good enough to directly observe planets around other stars, it is time to introduce the image of a planet around a star called beta Pictoris (beta Pic for short).[1] Why choose beta Pic to observe? There are a few reasons. First, it is a bright star. The system for correcting for atmospheric blurring needs a bright star to measure how the atmosphere is affecting the observation. This is because it needs to take many images of the star very quickly. This leads to very short exposures where only the brightest stars will appear. It is possible to use a fake star, a spot generated by shining a laser into the upper atmosphere, but it is easier just to have a bright star. Another reason is that beta Pic has been known since the 1980s to have a debris disc around it. Debris discs are the leftovers from the planet formation process described in the last chapter and are a sign that planets may exist around a star. The final reason is beta Pictoris's young

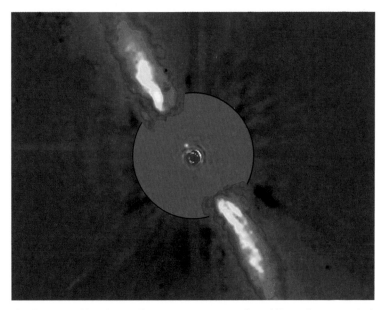

The planet around beta Pictoris. The image is a composite of two different observations, both of which removed the blurred, bright light from the parent star. The outer part shows light from the central star bouncing off the disc (seen edge-on here). The inner circle is from an observation with a better correction for blurring from the atmosphere. It shows a black circle in the centre surrounded by circular patterns and, to the upper left, a bright point. This is a young, recently formed planet.

age of about 24 million years.[2] That might not sound young, but it is in a galaxy that is more than 12 billion years old. And that young age turns out to be pretty handy.

Planets are like embers spat out by a fire. They form in hot environments and absorb the heat from the material impacting on their surfaces as they grow in discs. This means that, unlike a star, even though a planet has no internal nuclear furnace it still glows. Once the planet has formed it behaves like the aforementioned ember, cooling as it radiates heat away into its surroundings.

If you are looking to observe a planet directly, you want it to be as bright as possible compared to the star it is orbiting. The best way to maximize this brightness ratio is to catch the planet when it is at its youngest and therefore brightest.

What does the image show? It's actually the composite of two images, both taken in infrared light with adaptive optics systems. The outer image was taken from the European Southern Observatory's 3.6-metre (11.8 ft) telescope in La Silla, Chile. This composite picture was made in 1996 and shows light from beta Pictoris reflecting off the debris disc in orbit around it. The inner image is taken with a better adaptive optics system on the larger 8-metre (26 ft) Very Large Telescope in Paranal, also in Chile. The centre of the image is masked out to remove the brightest light coming from beta Pic. Around this blocking circle are a series of structures left over from the slightly imperfect removal of the pattern of spread-out light made by the telescope. The bright point just above and to the left of beta Pictoris is the planet beta Pic b, separated by about four-tenths of an arcsecond from its parent star. The light reaching us from beta Pic b mostly comes from it radiating away the heat from its formation. There's very little reflected light from the primary star. Note that it looks like the planet is roughly in line with the debris disc in the wider image.

WHEN WE LOOK up at the sky we see a three-dimensional universe projected into two dimensions. Under clear conditions the star Alpha Capricorni can be resolved with the naked eye into two stars very close to each other. You might guess that this is a binary star, a system with two stars orbiting each other. In fact one of the components is made up of three stars too close for the human eye to separate. All three lie one hundred light years from Earth. The other component is a star system almost seven hundred light years away. Binary stars are rarely separated by one light year, let alone six hundred. This means that Alpha Capricorni is made up of two unrelated star systems that by coincidence look close together.

How do we know that beta Pictoris has a planet around it and not just another star in the background that looks close to it but is actually a long way behind it? There are a couple of ways to do this. The first is for astronomers to say, 'Looking at maps of the sky we can work out how many random stars above a certain brightness there are in a big area of sky. So, we can work out the odds of finding a similar star in the small area close to our target star.' This is a bit like knowing how many planes land at an airport in an hour and dividing by sixty to work out the odds of one landing in a particular minute. If the odds of finding a background star in the area around beta Pictoris are really small then it is likely that the companion to the target is a planet, not an unrelated star that just appears to be close to it by coincidence.

The other way to check that you are not just seeing a background star is to observe the object several times over a number of years. Beta Pictoris is moving across the sky very slowly, and I mean *very* slowly. Imagine something moving across your field of view at the walking speed of a tortoise, but as far away from you as the Sun is from the Earth. This is imperceptibly slow, but still twelve times the rate beta Pictoris appears to move across the sky. A coincident background star will typically appear to move more slowly across the sky than a nearby star like beta Pictoris. If astronomers take a few images of the candidate planet a year or two later then they would see a real planet moving along with its star, rather than a background star moving through space with a different speed and direction.

Did astronomers see the motion expected for a planet described above with beta Pic? Six years after the original 2003 image, a subsequent observation showed the planet had moved with beta Pictoris, but had swapped to the other side of the star.[3] The astronomers had seen the planet moving in its orbit. Subsequent images picked up more of this orbital motion.

What else do we know about the planet shown in the image? The temperature at the top of its clouds is about 1,400°C (2,550°F), it is orbiting about twelve Earth–Sun distances from its parent star and it is a gas giant like Jupiter. It is about the same size as Jupiter but with a little under thirteen times as much mass.[4]

Before explaining how astronomers worked out the properties of beta Pic b, I should first introduce you to some very similar planets that were found at the same time.

12

AN EMBER SPAT FROM A FIRE

This chapter begins with a tired cliché. Not so tired that it should be avoided like the plague, but par for the course: you wait years for a directly imaged planet around another star, and then five come along at once.

As with other planet-detection techniques, directly taking a picture of a planet was the result of decades of technological improvement and careful development of data-reduction techniques. Successive generations of instruments produced better corrections for atmospheric blurring, allowing planets to be found closer to their parent star. The first directly imaged exoplanet was 2MASS 1207b, found in 2005.[1] The object this planet orbits, however, is not a star but a brown dwarf. This is a type of object that is even more massive than a gas giant planet like Jupiter, but not massive enough to have the nuclear furnace that powers a star like the Sun. We had to wait a few more years before the first planet around another star was imaged.

In November 2008, when the world's financial crisis was beginning to bite, three announcements were made of planets imaged around other stars. One was beta Pic b, another was a body that appeared to orbit the young star Fomalhaut.[2] This latter discovery is an odd object that is not visible in infrared light but can be seen with the Hubble Space Telescope in visible light. That is not what you would expect from a young planet, so either this is not a planet orbiting Fomalhaut or it is a planet enshrouded by a dust cloud, which reflects light from the primary star and makes it impossible to see the emission from the planet itself. Finally there was HR 8799, a young star of similar temperature to beta Pictoris.[3]

The search for planets around HR 8799 followed roughly the same process as the beta Pictoris study. HR 8799 was picked for observation because it is a young star discovered with a debris disc. An image of it was then taken using an adaptive optics system

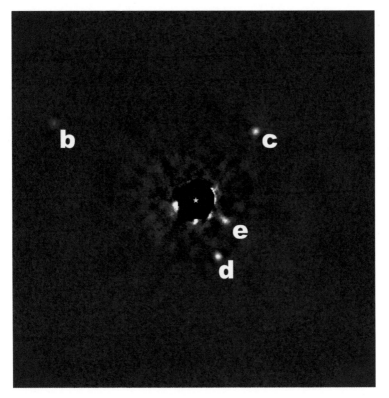

An image of the four planets orbiting the young star HR 8799. The features around the central star are a result of the process to remove the spread-out light caused by atmospheric blurring and telescope optics. Multiple observations over several years have shown that these planets orbit almost face-on in an anti-clockwise direction. The four planets have projected orbital separations of 14, 24, 38 and and 68 Earth–Sun distances respectively.

to remove atmospheric blurring. The astronomers studying HR 8799, however, employed another clever trick to further improve the quality of their images.

The images taken using adaptive optics systems are not flawless as imperfections in both the telescope optics and the process of correcting for atmospheric blurring can lead to blobs around the star known as speckles, which could easily be mistaken for a planet.

Modern telescopes track stars across the sky using motion in two directions, up and down and left and right (around in a circle on the ground). Unfortunately, the complexities of tracking a point on the sky mean that unless your telescope is at the North or South Pole, the image you take of the object you are interested in will rotate over time as the field you are observing moves during the night. Telescopes have optical systems built into them to remove this field rotation. If you turn these off, however, the field will rotate, but all the nasty speckles left over from the telescope optics will stay in the same place. This means that your planet will move with the field rotation, making it stand out from the static speckles. This technique allows astronomers to discover planets closer to their parent star than they otherwise would.

In 2008 three planets were discovered around HR 8799 using an adaptive optics system and the abovementioned clever series of tricks to improve images. Then, in 2010, further observations identified another planet, closer to the star than those discovered two years before.[4] Later observations have shown that these planets follow arc-shaped paths centred on HR 8799. This demonstrates that they are orbiting their parent star. As with beta Pic, we have a planet or planets that appear to orbit a bright star. But how do we know these are planets and not other types of objects orbiting their stars?

Cast your mind back to the last chapter and the comparison between a planet and a spark or ember spat out from a fire. A tiny spark may fizzle out almost immediately as it drifts through the cooler surrounding air. On the other hand, a more substantive ember may land on the ground and remain hot for some time. This leads to a problem. Say I tell you I found an ember that had been spat out of a fire and had managed to measure its temperature. How would you work out how big the ember was?

We have the same problem with directly imaged planets. Take HR 8799 b, the furthest-out planet in the system. Astronomers can take images of the planet in different coloured filters. The separate brightness measurements in the different filters can be compared to distinct computer models of giant planet atmospheres with varied temperatures. The result shows an object that appears to have a temperature of 1,000–1,500°C (1,830–2,730°F), depending on the atmosphere models used.[5] This temperature range is much cooler than the coldest-known type of star burning hydrogen in its nuclear furnace. But that does not quite make HR 8799 b a cast-iron planet.

When looking at the solar system it is pretty easy to say what is a planet and what is a star. The star is the big thing in the middle burning hydrogen; the planets (major, dwarf or minor) are the things going around it. The reason that this distinction is easy to make is that the gap in mass between the Sun and the second most massive thing in the solar system (Jupiter) is pretty big. Jupiter is about 1,000 times less massive than the Sun and has far too little mass to ever fuse hydrogen in its core. The problem is that there is an in-between class of objects called brown dwarfs. These can form from a collapsing cloud of gas in a similar process to star formation or can form as a companion to a star. They do not have enough mass, however, to fuse hydrogen in their core. The smallest amount of mass a star can have and still fuse hydrogen is about 8 per cent of the mass of the Sun. This provides a neat dividing line between stars and brown dwarfs. But there is another question, what separates a brown dwarf orbiting a star from a planet orbiting a star? Currently astronomers use a simple but imperfect dividing line. Above about thirteen times the mass of Jupiter an object can briefly fuse deuterium, a rare, heavier form of hydrogen. This mass limit is used as a common dividing line to differentiate planets, which do not fuse deuterium, from brown dwarfs, which do.

The nuclear furnace at the heart of a star gives it a stable energy source for tens of millions of years. Lower mass stars burn for even longer, tens of billions of years – meaning that, once a star turns on, it burns constantly. However, deuterium fusion contributes energy to a brown dwarf only briefly. This means that brown dwarfs are like planets, embers thrown from a fire, cooling forever. If you find something orbiting a star that is too cool to be a star itself, how can you tell if it is a planet or a brown dwarf?

Let us go back to the ember that has been spat out of a fire. Is it a smaller ember recently ejected from the flames, or a larger ember that flew out a while ago? We have a measurement of the cinder's temperature, now let us add an extra tool, a stopwatch. If we know the temperature and how long ago the ember came out of the fire, then we can work out how massive the ember is.

The same applies to an object that may be a planet or a brown dwarf. If we estimate the temperature and have a measurement for the age of the planetary system, then we can use computer models of how planets and brown dwarfs cool with time to estimate the mass. This will tell us if an object orbiting the target star is a planet or a brown dwarf.

So how can you get a stopwatch that can measure the age of a star? There are a few techniques that astronomers can use, some more accurate than others. Measuring the age of an individual star is pretty hard, but measuring the age of a star cluster is easier. Most young star clusters represent the product of a short-lived burst of star formation. This means all the stars in the cluster should be about the same age. Young stars are slightly puffed-up compared to older stars and have different temperatures and brightnesses to older stars of the same mass. As they get older they slowly deflate and begin to look more like an old star of the same mass. Stars of different masses deflate at different rates, so astronomers can plot the distribution of temperatures and brightnesses of stars in a cluster and compare them to models of star clusters at different ages. A model for the distribution of temperatures and brightnesses for a cluster of stars of a particular age is known as an isochrone (meaning same age). The best-fitting isochrone for the cluster will give the best age estimate for its constituent stars. Another way to measure the age of a cluster is to look at some of the ashes of the Big Bang.

The Big Bang produced a lot of heat and light but did not produce any elements other than hydrogen. In the few minutes afterwards, however, the whole universe was as hot as the interior of a star. This meant that the hydrogen burned into other elements. During this brief time a little over a quarter of the hydrogen in the universe was turned into helium. This also produced a small amount of lithium, an element now commonly used in batteries in many portable electronic devices. Lithium is also burned in the cores of stars. This happens so quickly that the core of a star can have all its lithium burned out in a few million years. Low mass stars have churning interiors, so as lithium in the core is burned, fresh lithium-rich material from the stars' outer layers is mixed in. Over time this process leads to all the lithium in the star being burned.

Once lithium burning turns on, a low mass star is quickly drained of this element. A more massive low mass star will start burning lithium earlier than the lowest mass star. This works like a clock. First the stars of about a third of the mass of the Sun lose their lithium and, as time goes on, lower and lower mass stars become lithium-depleted. Astronomers can look at a cluster of stars, take a spectrum of each star and then look to see which stars have the spectral signature of lithium and which do not. They can then spot the boundary between the type of stars in the cluster with lithium and type of stars without lithium. They then know that the cluster is older than the amount of time it would take to burn all the lithium in the stars which they have found to be lithium depleted. They also know that the cluster is younger than the amount of time it would take to burn all the lithium in the stars that still have their lithium.

You might think it would be easy to measure the age of HR 8799 and its planets, as it is a member of a star cluster, but unfortunately HR 8799 is not a member of a young star cluster. Luckily it is a member of another type of group of young stars.

Stars within a few hundred light years of the Sun move with essentially random motions relative to us. There are, however, groups of stars that, although spaced out by tens of light years, appear to be moving in the same direction. These moving groups are likely to be the remnants of sparse young star clusters that have dispersed. Like a young cluster, a moving group represents the output of a single star-formation event, so all member stars will have the same age.

Luckily HR 8799 is moving through space in the same direction as one of these moving groups, known as the Columba

moving group.[6] Beta Pictoris is such an important young star
that the moving group of which it is a member bears its name.

Using the distribution of temperatures and brightnesses of
its constituent stars, astronomers can work out that the Columba
moving group is about 42 million years old.[7] For context, all the
stars in this group formed after the era when triceratops and
tyrannosaurus roamed the Earth. This age gives all the HR 8799
planets masses of six to ten times the mass of Jupiter. The beta
Pictoris moving group has age measurements from both the distribution of temperatures and brightnesses of its component stars
and from the lithium depletion boundary. Both give an age of
roughly 24 million years, implying that beta Pic b has a mass that
is just below the 13 Jupiter mass boundary between a planet and
a brown dwarf, suggesting that it is a planet.[8]

The HR 8799 planets are all much more massive than Jupiter and
in wide orbits with projected separations ranging from fourteen
to 68 Earth–Sun distances. That is a problem for the mechanism
for forming planets that starts from dust grains sticking together
and works up to larger and larger bodies amalgamating. It is not
able to form large planets in orbits wider than about 35 Earth–Sun
distances.[9] An alternative mode of planet formation has been suggested for objects such as HR 8799 b. In this, eddies and vortices
in the disc around the parent star are so extreme that they can
collapse under their own gravity to form planets. Models indicate
that this process can form massive planets on wider orbits.

We have seen that groups of young stars moving through
space, and the techniques for measuring their ages, can provide a
stopwatch to measure the age of the stars and the planets that orbit
them. But these groups of stars can also reveal a strange, unex-
pected sort of planet, one that will challenge your perceptions
of what a planet is.

13

A LONELY PLANET

As autumn approaches, I trudge along to the doctor for the annual ritual of being poked in the upper arm with a needle. This jab protects me from what are expected to be the most common variants of flu in the upcoming northern hemisphere winter. The deactivated viruses in this injection are most likely to be grown in hens' eggs. What the vaccine does not contain is brisket, or rump, or flank, or tripe, or any other part of a cow (*vacca* in Latin). Despite this, what protects me from the coughing, sneezing and aching joints is known as a flu vaccination, a name deriving from the Latin word for cow. This is because the first vaccination developed against smallpox used material from cases of a similar disease, cowpox.

The word 'planet' derives from the ancient Greek word *planētēs*, meaning 'to wander'. This comes from planets wandering across the sky due to their orbits around the Sun compared to the 'fixed' stars. But what if something that looks like a planet was found that was not wandering around a star. Is that still a planet just as my egg-grown flu vaccination is still a vaccine?

Let us go back to moving groups, the loose associations of young stars drifting through the region of space close to our solar system. Finding out if a star is a member of a moving group allows astronomers to determine its age. From this they can estimate the mass of any object in orbit around it and determine if it is a planet or a brown dwarf. If astronomers want to know if an object is a member of a moving group they need to know its position in space and the speed and direction it is moving in.

Typically, the process begins by finding lists of possible young stars based on certain signatures of youth. One such signature might be a debris disc around a star, similar to the ones around HR 8799 and beta Pictoris. Other signs of youth come from the fact that most stars are born spinning fast and begin to spin more slowly as they get older. How fast a star is spinning can be directly measured by looking at it for days on end and tracking how its brightness changes as light and dark patches on its surface come in and out of view. Faster-rotating stars also have more violent motions within them as the plasma (ionized gas) they contain churns and shears. This drives more powerful magnetic dynamos, meaning these stars have stronger X-ray emissions and suffer from magnetic explosions on their surface. Detecting signs like these is normally enough for an object to be placed on a list of potentially young stars. Once astronomers have a list of young stars they need to determine four bits of information in order to work out if any of these stars are members of a moving group.

Some types of astronomical observation are easy to do, some are hard and cost a lot of telescope time. One of the easiest observations is to find the motion of a star or brown dwarf across the sky. This normally consists of taking images of the same patch of sky a few years apart and looking for changes in the positions of objects. This, along with a star's position in the sky, gives you the first two bits of information to determine whether that star is a moving group member. Why are these measurements useful? Imagine being in a field as a gargantuan flock of migrating birds heading towards the same destination flies all around you: some to your left, to your right, some higher up, some closer to the ground. Some birds that pass close to you might zoom past quickly, while others that are further away might appear to move more slowly. In reality they are all moving at roughly the same speed and towards the same point on the horizon. The apparent motions of moving group members are similar to that flock of birds. They will all move towards the same point on the sky because they are all moving with the same speed and direction. Despite these birds all moving at roughly the same speed, the closer ones will appear to move faster than those that are further away. Think of standing on the pavement and having to quickly turn your head to follow a car driving past versus your eyes tracking a car moving at the same speed along a road on a distant hillside. The solar system is surrounded by many of these flocks of young stars as well as thousands of other stars all moving in what appear to be random directions.

If an astronomer finds an apparently young star that is moving towards the same point on the sky that members of that moving group are heading for, how do they work out if this is a moving group member and not just a random star with the same apparent motion across the sky? Having done the easy observation looking for motion across the sky, the astronomer now has to get two more bits of information using more costly techniques.

Cast your mind back to some of the first alien worlds mentioned earlier that were discovered by monitoring the radial velocity (the motion towards or away from us) of their parent stars. Using the same technique, but typically one or two observations rather than several tens, our astronomer can measure the star's motion towards or away from the Earth. This is the third bit of information.

There now follows a short interactive section. Hold your index finger in front of your face and close one eye. Now close your open eye and open the closed one. Notice how your finger

shifts compared to the background? Now move your finger further away and repeat the process. You will notice the shift gets smaller the further away you move your finger. A similar effect is used to measure the distances to stars; take images of a star over a year and the changing perspective as the Earth moves around the Sun makes the star move a little compared to stars in the background. The further away the star, the smaller this shift in position will be. The size of this shift can be used to measure the distance to the star. This distance is the fourth and final bit of information that is needed.

Once our astronomer has all four bits of information – the position of the star on the sky, the motion of the star across the sky, how fast it is moving towards or away from the Earth and how far away it is – they can describe the star's position and motion in three-dimensional space. If this position and motion matches the expected distribution of positions and motions of a known moving group of stars, then it is likely the astronomer has just found a new moving group member.

THIS BRINGS us to PSO J318.5-22, a magenta dot first identified by astronomers using a survey telescope called Pan-STARRS1.[1] They initially flagged it as a possible brown dwarf that was unusually red. We met brown dwarfs in the last chapter as objects more massive than planets that orbit stars. Brown dwarfs can also float through space on their own, just like stars. PSO J318.5-22's curious colour drew an interesting comparison with the planets around HR 8799. As mentioned in the last chapter, HR 8799 b has a temperature in the 1,000–1,500°C range. There are a number of brown dwarfs with similar temperatures, but HR 8799 b appears much redder than them. A number of possible reasons have been suggested for this, including that HR 8799 b has very thick clouds that block out a lot of the blue light coming from the planet's interior but let the red light through.

This similarity to HR 8799 b changed PSO J318.5-22 from being a relatively nondescript brown dwarf destined for an anonymous entry in a large catalogue of objects to something more interesting, a possible young planet. Astronomers measured the position, the motion of the star across the sky, the line of sight velocity and the distance. These all show that PSO J318.5-22 has the same motion through space as the beta Pictoris moving group. This means it is likely to have the same common origin, and hence age, as this

The red dot in this image is the free-floating planet PSO J318.5-22. This picture is made from multiple images taken in different colours of light for the Pan-STARRS1 survey, the survey which first identified PSO J318.5-22.

group. Using this age of about 24 million years and the observed brightness, colour and distance of PSO J318.5-22, it is possible to work out that it is about 8.5 times the mass of Jupiter.[2] This puts it well below the deuterium-burning boundary, the upper limit for a planet's mass. But it is a planet floating through space on its own. How did this lonely planet come to be drifting through the Galaxy with no parent star?

As mentioned in the last chapter, in November 2008 five directly imaged planets were announced within a few weeks of each other. PSO J318.5-22 had its own little bit of astronomical serendipity. On the same day as the announcement of its discovery, a study on an object in a cluster of young stars was published. This was OTS 44, an object twelve times the mass of Jupiter that is only a few million years old and like PSO J318.5-22 is not orbiting a parent star.[3] Notice the mass: this puts it just below the planet/ brown dwarf boundary.

As mentioned earlier, stars form from collapsing clumps of gas that spin faster as they get smaller. This results in a star in the centre and a disc of material surrounding it. Early in the star's history it can draw material from the disc onto its surface.

The study of OTS 44 showed two things. The first was that OTS 44 had a lot of warm material in orbit around it. This suggested that, like a young star, it is surrounded by a disc of gas and dust. The study also looked at the spectrum of light from OTS 44 and found signatures suggesting it was drawing that material onto its surface. Put together, this implies that OTS 44 has a disc of gas and dust around it and that it is drawing material from that disc onto its surface, just like a young star.

Here we have an object that has a similar mass to PSO J318.5-22 and appears to be behaving like a newly formed star. Therefore, it is perfectly possible that PSO J318.5-22, despite having the mass of a planet, formed like a star. It is also possible that the collapsing eddies and vortices in discs like those around HR 8799 could form an object that is too massive to be a planet and is instead a brown dwarf.

PSO J318.5-22 is an object that probably formed like a star and does not wander around a stellar parent but has similar physical characteristics to a planet. Meanwhile, processes that can form planets can also form things that are not planets. This shows that the deuterium-burning division between planets and brown dwarfs is not that meaningful.

It is also difficult to tell the difference between a five Jupiter-mass object that formed from gobbling up pebbles in a disc around a young star (like the planets forming around HL Tau) from one that formed from a collapsing eddy in a disc (the way that the planets around HR 8799 probably formed). A planet formed by either of these processes could get a huge gravitational kick from another planet in orbit around its parent star and be thrown out of its orbit and into interstellar space.

An object below the deuterium-burning limit, floating alone in space, could have been formed by any of three processes: one star-like, two planet-like. If we cannot differentiate between objects formed by these three processes then, alas, we have no better dividing line than the pretty meaningless deuterium-burning limit to separate planets from brown dwarfs.

IT IS GREAT to find a sneaky way to make something easier, like shaving a couple of minutes off your walk to work by finding an alleyway to dart down. PSO J318.5-22 is a sneaky shortcut for astronomers. When studying planets like the ones around beta Pictoris and HR 8799, observers first have to spend a huge amount of effort getting rid of the light from the planets' parent stars. This isn't an issue for PSO J318.5-22 as there is no parent star to do away with.

Being an object with no annoying light from a primary star made PSO J318.5-22 an ideal candidate for one kind of observation: monitoring its brightness over several hours. What astronomers found when they did this was that PSO J318.5-22 was changing in brightness as it spun around on its axis of rotation.[4] This shows that PSO J318.5-22 is not uniform: brighter and darker patches move in and out of view as PSO J318.5-22 rotates, making it appear to get brighter or darker over time. For stars such a result would typically mean that it had starspots, darker patches caused by magnetic fields. PSO J318.5-22 is too cool to have starspots, so something else is causing the variability.

What is giving PSO J318.5-22 these bright and dark patches? The best guess we have at the moment is that there are patchy clouds in the atmosphere of this lonely planet. Where there are gaps in these clouds we can see deeper into hotter parts of the atmosphere. But these are not the ordinary water clouds you might see if you looked out of the window right now.

14

A DREICH WORLD?

Being Scottish, my speech is often peppered with Scots words that have seeped into Scottish English. A great example of this is 'dreich', a word originating from the Norse for 'enduring', which accurately describes a cloudy, cold, wet day in Scotland.[1] Miserable, unrelenting drizzle is, though, but a minor inconvenience compared to molten iron rain and clouds made of vaporized rocks.

The descriptions of the actual planets mentioned in the last three chapters have been a bit vague. They've been depicted as roughly the size of Jupiter but several times more massive. Then there have been descriptions of the temperature at the top of their atmospheres as being a bit over 1,000°C (1,832°F) and some brief references to clouds. What do those last couple of references actually mean?

Planets like the hot Jupiters in the first five chapters and HR 8799 b, beta Pic b and PSO J318.5-22 are all thought to have clouds. As with so much of astronomy, we cannot fly to these planets and take images of swirling patterns of storms and hurricanes as we can with solar system planets. Instead we rely on theoretical models of the atmospheres of planets and brown dwarfs, testing how well they compare with observations taken by telescopes. The models that currently provide the best match to observations of brown dwarfs and directly observed giant planets around other stars include clouds, but not as we know them on Earth.

Clouds in our atmosphere form when the water vapour in the air gets to a temperature and pressure where molecules can stick together to form relatively big droplets that can survive for a significant amount of time without being ripped apart by collisions with other droplets. These droplets are big enough that they block and scatter light. This means that the water vapour becomes opaque, allowing the clouds to be seen, be they white and wispy or dark and threatening.

The clouds astronomers think exist on planets like HR 8799 b, beta Pic b and PSO J318.5-22 are not like the water clouds we are used to seeing on Earth. As the temperatures at the top of their atmospheres range from 1,000–1,500°C, there is nowhere in the atmosphere of these planets where the temperature and pressure are low enough to form water clouds. These higher temperatures mean there is a much wider range for vaporized material in the atmospheres of these planets.

If the planets from the last three chapters do not have water clouds, what do they have? The clouds highest up in their atmospheres are thought to be made of vaporized iron and forsterite (a silicate mineral common in the Earth's crust). Above these clouds the atmosphere is cooler and the pressure lower. Here the iron and forsterite will cool to the point where droplets would become so large that they will rain out, vaporizing once more as they reach deeper, hotter parts of the atmosphere. Below these iron and forsterite clouds are thought to lie clouds made of perovskite (another mineral found in the Earth's crust) and corundum, the mineral that rubies and sapphires are made of.

In cooler giant planet atmospheres, the types of clouds mentioned above still exist, but only in the deeper, hotter parts of the planet. Higher up there are clouds made of sulphides and chlorides (such as clouds of vaporized potassium chloride, the sodium chloride substitute found in low-sodium salt). The upper layers of the atmosphere may have water or even ammonia clouds. Both are seen in Jupiter. Can we directly observe a planet outside our solar system that has water clouds like the Earth and Jupiter?

This chapter's planet, WISE 0855-0714, is floating through space alone like PSO J318.5-22,[2] although unlike that planet it is not a member of a well-defined group of stars. It is, however, so cold that even if it was one of the first objects to form in the disc of our Galaxy 10 billion years ago it would still be only ten times the mass of Jupiter. This means it has a mass below the deuterium-burning dividing line between a planet and a brown dwarf. Unlike most of the planets we have met so far, WISE 0855-0714 is pretty close to the Earth. Actually, it's the fourth closest system to the Sun after Alpha Centauri, Barnard's Star and a binary brown dwarf known as Luhman 16.

So far I have talked a lot about the clouds that are thought to exist in the atmospheres of young giant planets. But the gas that lies in-between and above them is also important. For planets and brown dwarfs colder than about 1,100°C (2,012°F) methane can exist in the upper atmosphere. This is the same substance used for cooking on a gas hob. Methane has a big effect on the observed properties of a planet because, like most molecules, it is a greedy eater.

As mentioned in Chapter One, atoms are picky eaters, only nibbling photons at specific energies. Atoms in the atmosphere of a planet will show up as thin lines, tiny bites taken out of the planet's spectrum. Molecules are different; they can be kicked by a photon in many different ways. Like an atom, a photon hitting a molecule can move an electron to a different energy level, but a photon can also make a molecule rotate and vibrate. This means that there are many more energies at which a molecule can absorb a photon. Rather than being picky, molecules chomp out huge bands from the spectrum of a planet.

Planets like beta Pic b, HR 8799 b and PSO J318.5-22 have water vapour (but not water clouds) and carbon monoxide in their atmospheres. Such molecules take big chunks out of the infrared light emitted by these planets and drastically alter the observed spectra astronomers see. This process is one-half of the greenhouse effect we see on the Earth. In our atmosphere there are relatively few molecules that take huge bites of the spectrum of visible light. This means that the visible light from the Sun passes through and hits the Earth's surface, warming it up. The Earth then radiates this heat back out towards space. Because of the temperature of the Earth, most of this radiation is in the infrared. The water, methane and carbon dioxide in the Earth's atmosphere gobble huge chunks of the infrared light heading out into space. This traps much of the heat in the Earth's atmosphere. The greenhouse effect keeps our planet much warmer than it would be if there was no atmosphere, so the greenhouse gases are not in themselves a bad thing. Global warming is a problem because humans are adding more greenhouse gases to the atmosphere, making the planet warmer than we have been used to. This is changing our settlement patterns and destroying habitats.

Let us get back to WISE 0855-0714. Once a planet's atmosphere gets cooler it is possible for methane to exist in its upper atmosphere. This methane joins the queue for the buffet, gobbling its own part of the planet's spectrum.

Brown dwarfs and planets colder than about 1,100°C have enough methane in their atmospheres to have a big effect on their spectra. These objects are also cold enough that, with no interference with molecules in their atmospheres, they would emit a huge fraction of their radiation in a portion of the electromagnetic spectrum known as the mid-infrared. At such wavelengths, methane is rather gluttonous, specifically for light with a wavelength of about 3.4 millionths of a metre. However, methane and the other molecules in the atmosphere of these brown dwarfs turn their noses up when presented with light with a wavelength closer to 4.5 millionths of a metre, so this light can escape into space. If you want to find an object cooler than 1,100°C, look for something that is bright at 4.5 millionths of a metre but not at 3.4 millionths of a metre.

Unfortunately, that can be a bit tricky as the Earth glows quite a lot at these wavelengths, so both the telescope astronomers would be observing with and the atmosphere they would be looking through would be relatively bright. There are ways to mitigate this on the ground, but there is also a place where these problems can be avoided: space.

The WISE satellite was launched in late 2009 with the mission of mapping the whole sky in the mid-infrared. Part of its mission was to look at light at several wavelengths, including around 3.4 and 4.5 millionths of a metre.

This brings us to the image of WISE 0855-0714. There are actually two images shown, taken seven months apart by the Spitzer Space Telescope shortly after WISE 0855-0714 had been discovered by the satellite. WISE 0855-0714 has moved between the two images. Remember how beta Pictoris was moving across the sky at about one-twelfth of the motion you would observe if you saw a tortoise moving across your field of view if it was as far away from you as the Sun is from the Earth. This is one-twelfth of an arcsecond per year, a unit astronomers use to measure motion across the sky. WISE 0855-0714 is moving at eight arcseconds per year. That is not because it is moving through the Galaxy

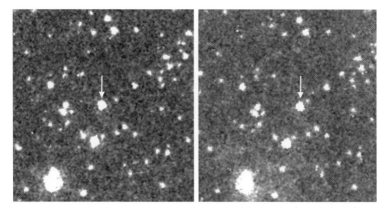

Two follow-up images of WISE 0855-0714 taken by the Spitzer Space Telescope, in June 2013 (left) and January 2014 (right). Both are taken in the mid-infrared, at a wavelength of about 4.5 millionths of a metre. Each image has a size one-fifteenth of the full Moon. After just over half a year, the space motion of WISE 0855-0714 is apparent.

particularly fast; it is because it is nearby, the fourth closest system to the Sun.

WISE 0855-0714 also stands out because it emits a lot more light at 4.5 millionths of a metre than 3.4 millionths of a metre, about 25 times more. This, combined with the fact that it is really, really faint, indicates that it is pretty cold. Combining a distance measurement, the mid-infrared WISE data and near-infrared data from the Hubble Space Telescope, astronomers worked out that WISE 0855-0714 probably has a temperature in the −45°C (−49°F) to −20°C (−4°F) range.[3] That roughly spans the coldest temperature ever recorded in Moscow to the coldest temperature ever recorded in Paris.

At this temperature WISE 0855-0714 is only 100°C (212°F) hotter than Jupiter and should have almost all the types of cloud mentioned earlier. First the perovskite and corundum at the deepest cloud level, then the vaporized iron and forsterite clouds and then the sulphide and chloride clouds. One open question about WISE 0855-0714 is whether any other types of cloud lie above this last level. The planet is too hot to have ammonia clouds like Jupiter, but careful studies of WISE 0855-0714's spectrum around 4.5 millionths of a metre show it may have clouds made of water.[4] So it could be a dreich planet. Unfortunately, 'could' is as far as we can go at the moment. Current theoretical models both with and without water clouds do not perfectly reproduce the emission of WISE 0855-0714 at various wavelengths.[5]

Based on its temperature and brightness, WISE 0855-0714 would have a mass less than the deuterium-burning mass limit even if it were one of the oldest objects in the disc of our Galaxy. In fact, it might have a mass similar to PSO J318.5-22 and is also floating through space with no primary star. In a few billion years, after tens of orbits round the Galaxy, the lonely planet PSO J318.5-22 could become a dreich planet.

WISE 0855-0714 is a gas giant that is cold enough for water clouds to possibly exist in its atmosphere. However, if we want to search for a planet with conditions similar to the Earth, we need to look for more solid ground.

LIFE

15

A CURSED WORLD

observed an Earth-like planet last week. I was not at a world-class observatory in some remote, picturesque location, or even looking through a telescope. I was standing on the balcony of my flat looking west just after the Sun had set. A high-tech satellite in space did not discover the planet I saw, it was first observed by the earliest humans. You might have heard of it, it is called Venus.

Earlier we discussed the term 'Earth-like planet' during the chapter on Kepler-10b (Chapter Nine). In some ways Venus is very similar to Earth. It is rocky and only a little smaller than our world. It has an atmosphere rich in oxygen and has a surface covered in mountains and valleys. Unfortunately for any would-be interplanetary tourist, Venus is very different from Earth in many other respects. The atmospheric pressure on Venus's surface is about ninety times that of Earth. Say you were sitting at the bottom of the Norwegian Trench, the deepest point in the North Sea, with the weight of a column of water 700 metres (2,300 ft) high pushing down on you. The pressure you would feel would still be less than the surface atmospheric pressure on Venus. The atmosphere on Venus is hellish: the oxygen in the air is in the form of carbon dioxide and the surface temperature is over 450°C (842°F).

So why am I telling you about our twisted twin world? Because the curse that has affected Venus has probably also affected the planet we are going to meet in this chapter.

THE FIRST SIX worlds we met earlier were discovered with telescopes that are small in comparison to the most advanced observatories astronomers have access to. We then met worlds discovered by the specialist Kepler satellite, the vast ALMA array of submillimetre antennae and some of the largest telescopes currently in existence.

The discovery of planets around other stars unleashed an outpouring of creative ideas from astronomers. Some resulted in expensive facilities like Kepler, others were cheaper projects like the array of small telescopes that discovered WASP-19b. The planet in this chapter was discovered with a new, small telescope built in an old observatory. TRAPPIST is a 60-centimetre (24 in.) telescope in La Silla in Chile, named after the beer-brewing religious order. It was built in a dome previously used for a small Swiss-owned telescope. This joint Belgian/Swiss instrument is automated and controlled remotely from Europe. Its mirror is around two hundred times smaller than those of the large telescopes that imaged beta Pic b and HR 8799 b.

The TRAPPIST telescope took measurements of the brightnesses of many faint, small, red stars known as red dwarfs. Stars of this type have a big advantage when it comes to looking for small planets. The depth of the dip in stellar brightness caused by a transiting planet depends on the size of the planet and the size of the star. An Earth-sized planet transiting a Sun-like star would produce a transit depth of 0.01 per cent. The same planet transiting a small red star would produce a transit depth of 1 per cent. This latter depth would be much easier to detect.

Alas that silver lining for planet-hunting astronomers has a cloud. Small red stars have very strong magnetic fields. Our own Sun has a magnetic field and the areas of the Sun's 'surface' that have the strongest magnetic fields are often populated with dark sunspots. These spots are slightly cooler than the surrounding areas of the Sun, so appear darker. The stronger magnetic fields on small red stars lead them to be very spotty indeed. They are covered with huge swathes of dark blotches. As the star rotates these darker patches come in and out of view. Sometimes there will be more dark blemishes on the part of the star visible from Earth, sometimes there will be fewer. This means that the brightness changes as the star rotates. Small red stars are very spotty so their brightness changes a lot over one stellar rotation period. This is not a killer blow for astronomers looking for planets around other stars; it just means they need to be careful to remove the extra variability caused by spots before looking for transiting planets.

In September 2015 the TRAPPIST telescope started observing a faint, red star situated about forty light years away from the Sun. In Chapter Twelve we came across the idea of a brown dwarf, something too massive to be a planet but too low in mass to have the nuclear furnace that powers a star. The particular star TRAPPIST was observing is only just massive enough to maintain nuclear fusion in its centre. It is just about as low in mass as a star can be.

Soon enough the astronomers operating TRAPPIST started to notice characteristic dips in the brightness of the star. Over time they observed more and more dips, evidence of one or more planets orbiting this star. After teasing out the signal left by the rotating spotty pattern on the star's 'surface', the astronomers were able to announce the discovery of three planets in the system.[1] These were:

TRAPPIST-1b, a planet that has a radius about 13 per cent bigger than the Earth with an orbital period of 1.5 days.[2]
TRAPPIST-1c, with a radius about 10 per cent bigger than

the Earth's and with an orbital period of 2.4 days. TRAPPIST-1d, about 21 per cent smaller in radius than the Earth.

The orbital period of the last planet was poorly defined at first due to times when the TRAPPIST telescope did not observe the star and thus missed some transits. TRAPPIST-1d was later found to be in a four-day orbit. Further observations from other telescopes were used to measure the masses of each planet using transit-timing variations. These showed that TRAPPIST-1b was about 1.02 Earth masses, TRAPPIST-1c was about 1.16 Earth masses and TRAPPIST-1d was only about 0.3 Earth masses.[3] These measurements, combined with the estimates of the size of each world, showed that all three planets are likely to be rocky. TRAPPIST-1c, the middle planet of these three, is the subject of this chapter.

The parent star of the TRAPPIST-1 system is small and low in mass. This latter factor, combined with the planets' short orbital periods, means that the system is incredibly compact. TRAPPIST-1b is separated from its star by about 1.7 million kilometres (1.1 million mi.). For comparison, Mercury, the closest planet to the Sun, is separated from its parent star by about 58 million kilometres (36 million mi.). A better comparison for the TRAPPIST-1 system in terms of scale would be Jupiter and its moons. The star TRAPPIST-1 is eighty times more massive than Jupiter but is roughly the same physical size. Jupiter has four large moons, all smaller than the TRAPPIST-1 planets. The outermost of these moons, Callisto, is separated from its parent planet by about 1.9 million kilometres (1.2 million mi.). As well as being similar in physical scale, the Jupiter moon system also shows a phenomenon that is relevant to the TRAPPIST-1 planets: the strange way the moons are kept hot.

When we think about what keeps the Earth warm we typically think of the Sun. There are, however, other sources of heat on the Earth and they are right beneath our feet. When our planet formed it was molten rock. The hungry young Earth in the early solar system gobbled gargantuan showers of pebbles. These slammed into the planet, generating heat. The Earth has since cooled and formed a solid crust, but the interior remains molten. This is partly due to the residual heat left within the Earth, even more than 4 billion years after its formation.

There is also another source of heat within the Earth. Radioactive elements like uranium are found in the Earth's core. These decay over time, producing heat. This helps keep our planet molten. The moons of Jupiter will also have formed as molten bodies and have radioactive decay heating their cores, but they have another energy source, too.

Way back when we met HD 209458 b, the second planet in this book, we came across the idea of tidal locking. HD 209458 b was stretched by the gravity of its parent star so that it had a bulge that faced the star. As the planet orbited, the star's gravity dragged on this bulge, forcing it to always face the star. The four large moons of Jupiter are also tidally locked. Jupiter's gravity stretches the moons and drags on their tidal bulges. This means the same side of each moon always faces Jupiter.

The stretching power of Jupiter's gravity also does something else. The moons pull and tug on each other and end up on elliptical rather than circular orbits. In an elliptical orbit the distance from Jupiter varies. As the tidal stretching forces on a moon vary with the distance to Jupiter, this means that the amount of tidal stretching changes throughout the orbit of each moon. Sometimes a particular moon will be close to Jupiter and will be stretched a lot, sometimes further away and stretched less. This constant changing in shape heats the moons up. Io, Jupiter's innermost large moon, feels the strongest tidal heating. This provides far more energy than residual heat from Io's formation or from internal radioactive decay. This heat could give Io a molten ocean of magma under its solid surface.

The TRAPPIST-1 planets are all tidally locked. Due to its short period and elliptical orbit TRAPPIST-1c will, like TRAPPIST-1b and TRAPPIST-1d, feel significant tidal heating caused by the stretching force of its host star's gravity. At the surface of each planet the heat flow from the interior caused by tidal stretching will be more than the heat flow the Earth has from radioactive decay and residual heat from formation combined.[4]

As well as the heat from tidal stretching, the TRAPPIST-1 planets are also warmed by light from their star. Small red stars give off less light than stars like the Sun, so planets in very close orbits around them need not be scorched worlds like Mercury or Kepler-10b. It is relatively easy to work out how much heat each TRAPPIST-1 planet would receive. When testing the habitability of each of these three planets you would think it would be simple. Do the sources of heat each planet receives result in a planet that has a temperature that could support liquid water? Alas, it is not that easy.

Radiation from the star at the centre of the TRAPPIST-1 system will hit each planet, warm it up and then each will radiate heat back into space. Like with WISE 0855-0714, however, the molecules in the atmosphere of each TRAPPIST-1 planet will take huge bites out of the spectrum of the outgoing light. This will trap energy in the planet's atmosphere, increasing the temperature.

This greenhouse effect warms up worlds with significant atmospheres. That does not mean the temperature of a planet just keeps increasing. The higher temperature means more heat

is radiated into space and hence a balance is eventually found between heat coming in and heat radiated out.

Venus in our own solar system has suffered a rather less benign greenhouse effect than the Earth. Our sibling world is likely to have once had an ocean early in its life. A few billion years ago the Sun was fainter than it is now, so Venus would have received less heat from the Sun than it does today. Eventually the Sun got brighter, Venus got warmer and the ocean started to evaporate. This evaporated water would have gobbled portions of the

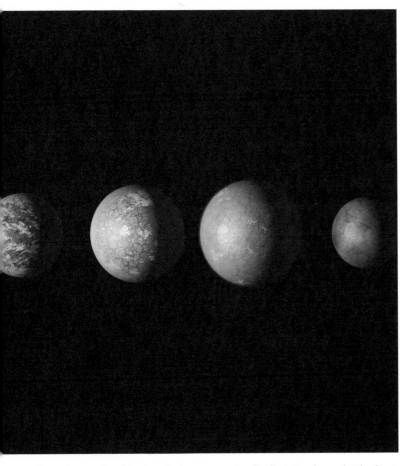

An artist's conception of the planets in the TRAPPIST-1 system. The illustration shows what the planets could possibly look like based on observations and models of the planets' climates. The planets and the star TRAPPIST-1 are shown to scale. The distance between the planets is, however, not to scale.

outgoing radiation, increasing Venus's greenhouse effect and warming it further. This warming would cause more evaporation, leading to more atmospheric water vapour and even more warming. The hot atmosphere would have strong convection currents dragging the water vapour into the upper atmosphere. There the water could be broken into its constituent elements of hydrogen and oxygen by ultraviolet radiation from the Sun. The hydrogen then escaped into space.

On the Earth water plays a regulating role in the atmosphere. Carbon dioxide dissolves in it and falls to the ground as slightly acidic rain. That rain can then react chemically with the rock, taking the carbon dioxide out of the atmosphere. On Venus, with the water destroyed, this regulating factor disappeared. Volcanoes pumped more and more carbon dioxide into the atmosphere, but there was no water to remove this greenhouse gas. This runaway greenhouse effect heated Venus to its current temperatures. It is far too hot and dry for life.

TRAPPIST-1c is likely to have suffered the same fate as Venus.[5] If it started with an ocean this would have quickly evaporated. The dayside of the planet would have developed huge convection cells, pushing water into the upper atmosphere, where it would have been destroyed. TRAPPIST-1c lies too close to its star to host a stable, temperate climate. If it does have a substantial atmosphere it has probably suffered from the same curse as Venus and is a hot, dry, dead world.

In this chapter we have met three planets orbiting the star TRAPPIST-1, but that was not the end of the story. Astronomers kept observing the star and saw more transits. This led them to identify more planets in the system, including the world we will meet in the next chapter.

16

A WORLD THAT'S JUST RIGHT?

At the start of this book we discussed how different the Earth was from the range of planets around other stars. We also discussed how different other planetary systems were from our own solar system. In this chapter we are going to meet a world that may be temperate and hospitable to life, just like the Earth is. But while it may seem familiar at first, it is found in a completely different astrophysical environment to the solar system.

In the previous chapter we met three planets in the TRAPPIST-1 system. These were roughly Earth-sized worlds in orbit around a star that is much cooler than the Sun and only 8 per cent of its mass. The discovery led to monitoring of the TRAPPIST-1 star using a variety of telescopes. Around a year after the announcement of those three planets, four more worlds were found in the TRAPPIST-1 system.[1] They were:

TRAPPIST-1e, a world with a radius 8 per cent smaller than the Earth's, in a six-day orbit around its star with a mass of 0.8 Earth masses.[2]

TRAPPIST-1f, a planet with a radius 5 per cent bigger than the Earth's, in a nine-day orbit with a mass of 0.9 Earth masses.

TRAPPIST-1g, a world with a radius 15 per cent bigger than the Earth in a twelve-day orbit with a mass of 1.2 Earth masses.

TRAPPIST-1h, a tiny world only 78 per cent of the Earth's radius that takes a little under nineteen days to orbit its parent star and with a mass barely a third of the Earth's.

All the masses above were measured using transit-timing variations, the delays and early arrivals of transits caused by the gravitational tugging of the planets on each other.[3]

WHILE THE fringes of science fiction contain jagged, hulking rock monsters and other life forms not based on carbon and water, most of our thinking about what life on other planets could look like revolves around a firmly Earth-based view of what a living creature is. To support life similar to what is found on our home world, another planet must be able to support liquid water.

Small red stars like TRAPPIST-1 give off less light than stars like the Sun. The habitable zone is a region around a star where a planet could be warmed to a temperature where liquid water could exist. As it emits relatively little light, TRAPPIST-1's habitable zone is much closer to it than it would be if it were a hotter star like the Sun. Despite orbiting close to their star in short orbits, it was possible some of the TRAPPIST-1 worlds may be able to support liquid water.

Astronomers have run computer models of the climates of three of the TRAPPIST-1 worlds – d, e and f – to test their habitability.[4] Planets g and h were not simulated as they are so cold that they cannot support liquid water on their surface. Simulations of TRAPPIST-1d showed that it probably suffered a runaway greenhouse effect similar to its neighbour TRAPPIST-1c. Thus it is likely too hot to have surface liquid water. Even with a thick atmosphere of carbon dioxide, TRAPPIST-1f was found to be too cold for liquid water on its surface. Thus it is likely a snowball world like its outer neighbours, planets g and h. This leaves TRAPPIST-1e, the subject of this chapter, a world that may fall in the habitable zone or 'Goldilocks zone', where the temperature is just right for liquid water.

The surface temperature of TRAPPIST-1e depends on its atmospheric composition. Too much carbon dioxide and it becomes too hot for most types of life found on Earth. It would still be able to support liquid water but it would be too hot for creatures with similar biology to humans. However, on the Earth there is a range of organisms that live in bizarre inhospitable environments. There are bacteria in the geysers of Yellowstone National Park in the USA, microorganisms in alkaline lakes in East Africa and strange creatures around volcanic vents deep beneath the ocean. Hence even if TRAPPIST-1e has an atmosphere that makes it too hot for you or me, it could still support extreme forms of life.

TRAPPIST-1e could, on the other hand, have an atmosphere with very few greenhouse gases. This would lead to water on its surface freezing. That is not quite the full story, though. All the TRAPPIST-1 planets are tidally locked to their star with one side of each planet constantly facing their stellar parent. This means that on an icy TRAPPIST-1e there would be one part of the planet where it was always midday with its star directly overhead. Around this warmer part of the planet the ice could melt, producing a small sea.

TRAPPIST-1e could also have just the right amount of greenhouse gases to maintain a surface temperature that could support liquid water and life. Again, the planet would have one point on its surface where it was always midday. Here it would be like midday in the tropics with huge clouds forming from evaporated water. There would also be one entire side of the planet where it was always night. However, the atmosphere would redistribute heat from the dayside of the planet, keeping much of the nightside free from ice. There would, however, be polar icecaps that are larger on the nightside.

These are three possibilities for TRAPPIST-1e. How could we possibly work out which of these three categories this world falls in to?

In Chapter Four we met WASP-19b, a hot Jupiter that had its atmosphere characterized by transit spectroscopy. A thin ring of atmosphere around the planet left its fingerprints on the planet's transit signal. Molecules in the planet's outer reaches munched at light from WASP-19b's parent star, changing the depth of the transit signal.

Transit spectroscopy would be a great way to characterize the atmosphere of a planet like TRAPPIST-1e. In fact it has been done already. The first such measurements from the Hubble Space Telescope have shown that the planet does not have an atmosphere rich in hydrogen.[5] This means its atmosphere is not like Neptune's, so perhaps it is like Earth's. What more could future transit spectroscopy measurements tell us about TRAPPIST-1e?

Let us take as an example a habitable planet that has been studied in detail, the Earth. Looking at pictures of the Earth from space we can see huge expanses of forests and grasslands. However, for most of the history of life on Earth complex organisms like large plants and animals did not exist. That does not mean you would not have been able to spot life from orbit. Microorganisms can produce gases like oxygen and methane. In fact they transformed our atmosphere, making it rich in oxygen and methane billions of years before plants or animals appeared. Such gases, and more complex molecules produced by biological processes, are known as biosignatures and could be detected by transit spectroscopy. It is worth noting that, as is so often the case, the universe can trick astronomers looking for these bio-signatures. For example, a desiccated planet that has suffered a runaway greenhouse effect could have lost the hydrogen from its water but held on to the oxygen.[6] Such a dead planet could look alive to astronomers studying it.

Unfortunately, TRAPPIST-1e orbits a rather troublesome star that makes transit spectroscopy tricky.[7] The star TRAPPIST-1 is a spotty, small, red star. This spottiness, caused by the star's strong magnetic field, makes the star vary in brightness as spots come in and out of view over a stellar rotation period. Transit spectros-copy measures how a star's spectrum changes when part of it is obscured by a planet's atmosphere. A planet transiting a spotty star will sometimes cross spots during a transit. Crossing in front of a dark spot will block less starlight than crossing in front of a bright patch of the star's 'surface'. This means that the amount of light blocked by the planet's atmosphere depends on whether it is passing in front of a spot or a brighter patch. Dark spots on the star are also cooler than the rest of the star's 'surface'. Cooler spots will have a different, redder spectrum than the rest of the star. Hence when the solid part of a planet and the surrounding atmosphere block out a dark spot it will have a different effect on both the total amount of light we receive from the star and on the spectrum of that light compared to if the planet were blocking out a bright spot.

All this means that the transit spectroscopy signal from the planet will jump around as it crosses different parts of the star. Add in the additional factor of spots coming in and out of view as the star rotates and you have a lot of noise shoved on top of an already rather tricky measurement. Recent developments in the modelling of these patchy host star 'surfaces' has led to some old results being revisited. Recall that in Chapter Four we met WASP-19b and saw that a recent study hadn't found some spectral features that were seen in other observations. It was more detailed modelling of the possibly patchy host star of WASP-19b that removed these spectral features. WASP-19b orbits a hotter, relatively unspotty star compared to TRAPPIST-1. It remains to be seen if transit spectroscopy of planets around stars like TRAPPIST-1 will be accurate enough to reliably detect biosignatures like methane and oxygen.

TRAPPIST-1e is a roughly Earth-sized planet that, depending on its atmospheric content, could host liquid water. It is, however, in a very different astrophysical environment from our world. It has one side that is in perpetual darkness and another side that is constantly warmed by its star. The TRAPPIST-1 system also differs from the solar system in another way: its size and architecture. There are seven terrestrial planets in orbit around TRAPPIST-1, compared to four in the solar system. The system is also small, being more like the Jupiter moon system in size. So how did such a small star get such a rich and compact planetary system?

The planets around the star TRAPPIST-1 probably formed in a manner similar to the still-coalescing worlds around HL Tau. A disc of material containing gas and tiny flecks of dust formed around the young TRAPPIST-1. Over time the dust clumped together into pebbles and then eventually planetary embryos.

In every planet-forming disc there is an invisible but crucial boundary, the snow line. Beyond this point the disc is cold enough for ice to form. A planet born inside the snow line will have relatively small amounts of water but may be enriched by icy bodies that formed further out colliding with it. This is one possible explanation for how the Earth got its water, from collisions with water-rich asteroids. A planet born outside the snow line can hoover up huge quantities of ice. This gives us a sneaky technique to study how the TRAPPIST-1 system formed.

In Chapter Eight we met the Kepler-36 system. This contains two planets, both a few times the mass of the Earth. Using transit-timing variations astronomers worked out the mass of each planet and used the depths of the transit signals to estimate planetary radii. They then built lots of computer models of each planet with different amounts of iron, rock, ice and light gases such as hydrogen and helium. The estimated planetary radii of each of these models were then tested against the observed radii of the two Kepler-36 planets. This constrained the composition of both worlds and showed that one was a super-Earth and the other a mini-Neptune.

Astronomers applied the same technique to the TRAPPIST-1 system.[8] The models showed that TRAPPIST-1c is relatively dry and that TRAPPIST-1b might have a small water envelope. The mass measurements for TRAPPIST-1d and TRAPPIST-1e were not accurate enough to get good constraints on their composition. The three outer worlds (TRAPPIST-1f, g and h) all showed signs that they had a lot of water. These three worlds are so far from their star that even with a thick atmosphere they will have icy surfaces. Tidal heating, however, may give them subsurface oceans similar to those on Jupiter's moons Europa and Callisto. But while Europa has an ice layer a few tens of kilometres thick above an ocean 100 kilometres (62 mi.) deep, TRAPPIST-1f is likely covered with an ice sheet 2,000 kilometres (1,240 mi.) thick.

The high water content of the outer three TRAPPIST-1 worlds gives us a clue about where they formed. All currently orbit closer to their star than the primordial snow line. This means that if they formed at their current orbits the disc around them would have been too hot to produce ice for planetary embryos to gobble up. So how did these three worlds get their ice?

If TRAPPIST-1f, g and h all formed much further away from their star they could have formed in a region rich in ice. Beyond the snow line there would have been lots of ice for them to gobble up. Hence, it is likely that the outer worlds in the TRAPPIST-1 system formed further away from their parent star and migrated inwards. This migration could either have been caused by interactions with the planet-forming disc or by interactions between planets. Eventually the TRAPPIST-1 planets arranged themselves into a series of compact and resonant orbits. For every orbit TRAPPIST-1h makes, g makes three orbits, f four, e six, d nine, c fifteen and b 24. This careful arrangement has persisted for more than 7 billion years.[9]

TRAPPIST-1e is an Earth-sized planet that could possibly support liquid water but is in a very different astrophysical environment to our solar system. Much of this situational difference is down to it orbiting a small, red star. But there are other small, red stars that produce much more hostile habitats for Earth-sized planets, and one of them is right next door.

17

A WORLD BOMBARDED

The Sphinx. I don't mean the iconic statue next to the pyramids at Giza, but an altogether different sort of carving. The Dunhuang Yardangs sit in the Gobi Desert, in Gansu in northwestern China. These monuments include statues called the Peacock, the Golden Lion and the Sphinx. This Sphinx is not a tacky reproduction of the great Egyptian monolith, in fact it is even older than its Saharan counterpart.

The Great Sphinx at Giza was carved by legions of skilled artisans from solid limestone bedrock. The Sphinx yardang at Dunhuang is also formed from solid rock, but it was created by a different sculptor, the desert wind. A yardang is an exposed rocky outcrop in a desert that has, over time, been eroded and shaped by wind and desert sand battering off its surface. These structures are typically aligned with the prevailing wind and can take forms that resemble animals or, in the case of the Sphinx yardang, a mythological creature, resembling the Sphinx at Giza.

In the last few chapters we have discussed how a star can affect the atmosphere on a planet and its chances of hosting life. In this chapter we're going to meet a world that could host life but which has, like the Sphinx yardang, been pummelled by ferocious winds.

The first planets discovered around other Sun-like stars were found using the radial velocity method. For much of this book we have focused on smaller worlds discovered by other methods, primarily transits. The radial velocity method, however, continued to discover planets with steady improvements in instrumentation and analysis, leading to smaller and smaller planets being identified with this technique as well.

Small red stars are spotty due to their high magnetic fields. As we have already seen, this causes problems when trying to detect transits of planets around such stars. These spots also cause problems for the radial velocity method.

Cast your mind back to Chapter Five, when we met HAT-P-7b, a hot Jupiter that seems to orbit its star in reverse. Astronomers were able to deduce the path of HAT-P-7b's orbit by looking at the effect it had on the spectral lines of its star during a transit. A star rotates so part of it is moving away from us, the observer, and another part is moving towards us. This means that the spectral lines in one-half of the star will be shifted to redder wavelengths and on the other half of the star they will be shifted to bluer wavelengths. When HAT-P-7b transited in front of the part of its parent that is rotating towards us it blocked out part of the bluer light for each spectral line. This made the lines redder so that it looked like the star was moving away from us. Similarly, when HAT-P-7b transited the redder part of its star this blocked red light in each spectral line and blue-shifted all of the lines. This made it look like the star was moving towards us.

Spots can cause a similar effect. This is because they are darker than the rest of the star so can also cause deficits in the light we receive in the blue or red part of a spectral line. A dark spot moving across the star's 'surface' as the star rotates will first cause a deficit in blue-shifted light, leading to it looking like the star is moving away from us. Then, later, once the spot has moved to the redder part of the star, there will be a deficit of red-shifted light, making it look like the star is moving towards us. Hence astronomers using radial velocities to find planets around small, red stars must be careful to take the stars' spottiness into account. If not, they could end up discovering a 'planet' that is just the occasional spot on the star's surface.

In practical terms the radial velocity method for planet discovery consists of making lots of measurements of the radial velocities of bright stars over periods of several months or years. Often the astronomers doing the observations do not find the signal of a planet in their data. As a result, there are lots of radial velocity measurements lying around in observatories' digital archives.

As statistical analysis techniques have improved, astronomers have gone back and examined old data. One group of astronomers based in Chile, the UK and other countries began looking again at archival data of Proxima Centauri, a small red star that is the closest stellar neighbour to our Sun. Remember how the solar system was like Edinburgh of the Seven Seas, huddled together on an island and separated from the nearest star by the vast oceanic expanse of space. Proxima Centauri is the nearest island. It is to the solar system what St Helena (the nearest populated island) is to Edinburgh of the Seven Seas.

The team of astronomers found a possible radial velocity signal from Proxima Centauri. Eager to confirm their result, the team took more observations using one of the telescopes in Chile that had taken the archival measurements they had used. These follow-up observations, combined with the archival data, showed a planet in an eleven-day orbit.[1] The nearest island to us in the vast interstellar sea has a town.

There have been no transit detections of this planet (named Proxima Centauri b or Proxima b for short), so we cannot measure its radius or orbital inclination. Hence we only have a minimum mass estimate, 1.3 times the mass of the Earth. Proxima Centauri is a small, red star so its habitable zone is much closer to what it would be if it were a star like the Sun. A planet in an eleven-day orbit sits right in the middle of Proxima Centauri's habitable zone. If Proxima b is close to its minimum measured mass then it is a rocky planet less than four light years away that could be capable of supporting liquid water on its surface. But there is a problem, its star.

The star Proxima Centauri is the lowest-mass member of a triple system with the other two components being the stars Alpha Centauri A and Alpha Centauri B. Proxima Centauri is about 15,000 Earth–Sun distances from these two stars.

As a low-mass, small, red star, Proxima Centauri is faint and cool. Stars like this have very strong magnetic fields. The magnetic fields around stars are like curled, taut springs. Sometimes some of the magnetic field lines uncoil themselves and send huge explosions of high-energy particles into space, causing the star to brighten. Our Sun regularly has relatively small magnetic explosions. The largest of these can lead to more spectacular aurorae near the poles on Earth. The most extreme solar magnetic explosions can damage power and communication grids. Indeed in 1859, when electrical communications networks were in their infancy, the largest solar storm on record caused telegraph equipment across the world to spark and set the telegraph paper alight.[2] Proxima Centauri, with its strong magnetic field, has much larger outbursts than the Sun. In one recent outburst this faint, red star, which is one hundred times too faint to see with the naked eye, brightened by a factor of 68.[3] Imagine being on Earth when the Sun suddenly got 68 times brighter.

Proxima Centauri also has a strong stellar wind. The flow of particles hitting Proxima b is more than ten times higher than

An artist's impression of Proxima b, a likely Earth-sized planet in orbit around the nearest star to the Sun. Proxima b's star bombards it with violent winds and ultraviolet light.

the solar wind hitting the Earth's atmosphere. Different parts of the star also blow out different wind intensities. This means that as Proxima b orbits around the star the intensity of the wind it is blasted with changes, with the peak being 2,000 times stronger than the solar wind the Earth receives.[4]

So how does this violent environment shape Proxima b? Is it like the yardangs of Dunhuang, stripped by the wind, or could it maintain an atmosphere and possibly life?

The Dunhuang Yardangs have had their rocky surfaces eroded by the wind. The rocky surface of Proxima b will not have been carved by the stellar wind, but the howling gale of particles unleashed by its parent star could have seriously impacted the planet's atmosphere.

Proxima b is not sitting unprotected in the path of the raging stellar wind, it probably has a shield. The Earth has a magnetic field that protects us from the stream of particles coming from the Sun. Proxima b may also have a magnetic field. This would protect the lower reaches of its atmosphere from the stellar wind. However, there is a weak point.

Near the magnetic poles of the planet the upper reaches of the atmosphere are less protected. This can lead to particles in this region of the atmosphere being ripped away from the planet by the stellar wind. During a stellar storm caused by a magnetic explosion on Proxima b's parent star the planet could have even more gas ripped out of its atmosphere at the magnetic poles. It has been calculated that it would take 365 million years for this process of erosion to remove an Earth-like atmosphere from Proxima b.[5] Given the planet is likely to be several billion years old, this would mean the atmosphere would have been stripped bare by now. Gases emitted by volcanoes on the surface of the planet, however, might be enough to balance the erosion and keep the atmosphere in place.

The stellar storms from Proxima b's parent star could also negatively impact on an important layer of its upper atmosphere. The Earth is protected from the extremes of the Sun's ultraviolet light emission by a layer of ozone (a form of oxygen) in our upper atmosphere. The storms of particles Proxima b has to endure could send waves of high-energy particles crashing into the planet's atmosphere, destroying its ozone layer.[6] This would make it harder for life to survive on the planet's surface.

It is not just stellar storms that threaten the atmosphere of Proxima b. Somewhat counterintuitively a cold, red star like Proxima Centauri emits a substantial proportion of its electromagnetic energy at the bluest wavelengths (ultraviolet and X-rays). This is due to the hot, diffuse outer layers of its atmosphere being heated by acoustic waves and its strong magnetic field. As a result,

Proxima b receives thirty times more ultraviolet and X-ray radiation than the Earth, causing havoc with any atmosphere the planet may have.[7]

In Chapter Fifteen we discussed TRAPPIST-1c and Venus, the latter of which likely had an ocean that heated and completely evaporated. Ultraviolet radiation from the Sun then broke the molecules in the water vapour into hydrogen and oxygen with the hydrogen floating away. TRAPPIST-1c may also have suffered from this runaway greenhouse effect.

Even if Proxima b has avoided a runaway greenhouse effect and hung on to some of its water, a similar process may still have affected it. The ultraviolet radiation will break up water molecules in its upper atmosphere with the hydrogen drifting away into space. This could lead to huge quantities of water being lost by the planet. These could range from less than the total amount of water found on the Earth to many times more. The amount of water lost to ultraviolet radiation, and with it the habitability of Proxima b, depends on the early history of the planet and its star.

In the last chapter we saw that if TRAPPIST-1e was formed further away from its parent star then it could be water-rich. This is because planets that form further out from their parent star, beyond the snow line, would be able to gobble up lots of chunks of ice during their initial growth.

Forming further away from a parent star also keeps a planet at a distance during an awkward stellar adolescence. TRAPPIST-1e and Proxima b both have parent stars that are cool, faint and low in mass. Lower-mass stars live longer than higher-mass stars. While some high-mass stars might live for a few million years before exhausting their nuclear fuel source, the lowest-mass stars could live for trillions of years. These low-mass stars also evolve more slowly. This means that, even after they are born and form planets around them, they are not behaving like a mature star. Stars are born shining brightly and then settle down to a comfortable middle age burning hydrogen steadily and shining at a constant luminosity. Low-mass stars can have a wild youth that lasts for several hundred million years.

If Proxima b formed at its current distance from its host star then it bore the brunt of a stroppy stellar adolescence. As the young Proxima Centauri was much brighter than it is now, Proxima b would have been much hotter early in its life. This would have driven a runaway greenhouse effect, evaporating oceans and driving the convection of water vapour to the upper atmosphere where it could be destroyed by ultraviolet radiation.

Hence for Proxima b to be habitable it would need to have formed at a distance from its host star where it would have avoided its star's extended bright youth and the resulting runaway greenhouse effect.[8] Forming further out would have allowed it to accrete enough ice to be water-rich. This would mean it could afford to lose some water to ultraviolet radiation and still hold on to an ocean. It could then have migrated to its current orbit once the star had settled into a steady, sedate middle age.

We have seen the ways a star can carve away and scar the atmosphere and oceans of a planet and affect its suitability to host life. But how else could a star damage a planet?

DEATH

18

A DARK CLOAK OF DEATH

Walking up a mountain you might notice hollows, places where rocks, water or (especially in Scotland) sheep droppings might collect. Larger hollows might be big enough to have small lakes known as tarns or corrie lochans. If you were to drop a pebble on some randomly chosen part of the mountain it would likely roll down into the valley below. Drop a pebble into one of these hollows and it will just roll to the bottom.

The same applies to the solar system. Drop a pebble at some random place in the solar system and it will probably start falling towards the Sun, like a pebble rolling into a valley. Drop a pebble close to the Earth and it will fall towards the Earth, like a pebble rolling into the hollow. This is because there is a region of space around the Earth that is dominated by the Earth's gravity. In terms of the solar system it is like a small hollow on the side of the big valley with the Sun at the bottom.

The Moon lives in the region of space where the Earth's gravity is stronger than the Sun's. It is like a pebble, rolling around in a hollow on the side of a big valley. If the Moon were further away from the Earth it would be like the pebble was rolling closer to the rim of the hollow. If the Moon were more than 1.5 million kilometres (932,000 mi.) away from the Earth rather than 380,000 kilometres (236,000 mi.) away, then its orbit would become unstable and it would eventually wander off away from the Earth and into the solar system at large. This would be like moving the pebble over the rim of the hollow and having it roll down the hill into the valley.

This means that, like a hollow on a hillside bounded by a rim, there is a natural limit about 1.5 million kilometres from Earth within which the Earth's gravity dominates over the Sun's.

THIS CHAPTER'S PLANET, WASP-12b, is, like a few of the earlier planets we met, a hot Jupiter, a giant planet in a very close orbit around its parent star. WASP-12b is 40 per cent more massive than Jupiter and is in an orbit with a radius of about one-fortieth of an Earth–Sun distance.[1] It takes a little over one day to complete its orbit around a star that is a little more massive and a little hotter than the Sun. WASP-12b and its star are about a thousand light years from the solar system.

WASP-12b was discovered because it transits its star, causing it to dim periodically as it is partially obscured. By analysing how much its parent star dims by, astronomers can work out the radius of WASP-12b – and it is big, so big that it is falling apart.

Observations of WASP-12b showed that it is puffed up like many other hot Jupiters. It has a radius about twice that of Jupiter but the outer reaches of its atmosphere extend out to three Jupiter radii from the planet's centre.[2] The sheer size of that creates a problem. Because it is so close to its parent star, the outer reaches of WASP-12b are far enough from the planet's centre that they feel a bigger gravitational pull from the planet's parent star than from the planet.

WASP-12b, like the Earth in the solar system, has its own little hollow on the side of the big hill with its parent star at the bottom. Rather than thinking about pebbles in the hollow, think of WASP-12b itself as water in the hollow. The atmosphere of WASP-12b is so extended that it has filled up its gravitational hollow and is

A lake or tarn in a hollow on the side of the mountain Helvellyn, in England's Lake District. The water sits in the hollow with a small stream at the far end overflowing into the valley below.

overflowing it. This is like a lake in a hollow on the side of a mountain. WASP-12b is so big that its outer atmosphere is overflowing its little gravitational hollow and flooding the space around its star with gas. This means that it is likely to be slowly losing mass as its outer gases are pulled away by the gravity of its parent star.

Because of its closeness to its parent star, the gravitational hollow WASP-12b makes is not exactly spherical but shaped like a teardrop. The point of the teardrop is aligned towards the parent star. It is likely that a large amount of the material that escapes WASP-12b's atmosphere flows through the point of this teardrop towards its star. This is like the stream seen flowing out the lake on the side of Helvellyn.

In the last chapter we met Proxima b and its active parent star. The magnetic field of that star led to colder spots in the region of the star's atmosphere, where most of the light comes from, and to knots and loops of hot gas in its diffuse outer atmosphere. As well as increased emission from ultraviolet light and huge quantities of hot plasma being blasted out in gigantic magnetic explosions, activity can also be observed by a star giving off light at very specific wavelengths. This is caused by emission from very hot vaporized metals like iron and magnesium. All but the hottest and most massive stars have some amount of stellar activity and hence all have this emission of light at these very specific wavelengths.

Well, almost all . . .

When astronomers tried to measure the stellar activity of WASP-12b's parent star they found something odd. Rather than detecting the characteristic emissions at very specific wavelengths, they found, well, not much. In fact WASP-12b's parent star had the lowest emissions at these wavelengths ever observed in any star of roughly the same mass and temperature. Either this was a star with incredibly low amounts of stellar activity or something else was going on.

So what else could be causing the really low emissions at these specific wavelengths? The obvious thing to look at is what makes WASP-12b's parent star unusual: the fact that it has a massive planet that it is currently ripping material from. Huge amounts of gas are being grabbed from the atmosphere of WASP-12b and being drawn towards the star. This material is not perfectly transparent and will absorb light at specific wavelengths. This gas contains the same vaporized metals that cause the emission at special wavelengths in the star. However, this gas is colder than the outer layers of the star, so it absorbs at these specific wavelengths. This covers up any emission from the activity of the parent star.

WASP-12b, then, is a planet enshrouding its parent star in a dark cloak of death, hiding its stellar activity. It is possible that what is happening to WASP-12b also happens to other planets

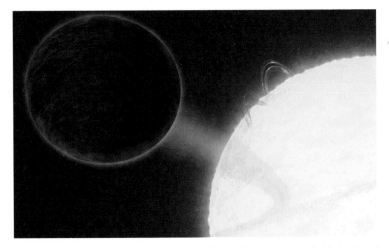

An artist's impression of WASP-12b shedding a stream of material onto its star. Here the black colour of the planet comes from a study which showed that WASP-12b reflects very little light from its parent star.

that wander into close-in orbits around their star. WASP-12b is probably massive enough that it could continue to exist for billions of years while slowly losing a very small fraction of its mass.[3] This would be longer than the remaining lifetime of its parent star. Smaller planets with smaller gravitational hollows would lose more of their mass due to both the blasting radiation (like Kepler-36b) and gravitational hollow overflow.

That's not the end of the story for WASP-12b. Recall how tidal forces moved HAT-P-7b closer to its parent star and at the same time started to turn its star around so it spun in the same direction as the planet orbited. The same process is gradually dragging WASP-12b closer to its parent star. Its orbital period is now about four minutes less than when it was discovered and it is likely that in 3 million years or so it will fall into its parent star and meet a fiery end.[4]

In this chapter we have met a star that is slowly killing its planet. But what happens when the star itself dies?

19

A WORLD TORN ASUNDER

In 1917, when much of the world was caught in the grinding horror of the First World War, a Dutch astronomer was making observations in California. He was looking at stars that moved quickly across the sky and he took a spectrum of one of them. Looking at the patterns of lines made by atoms nibbling away at the star's light, he reasoned it was probably an F-type star, a bit hotter than the Sun. It would take another ninety years before the significance of what he had found became apparent.[1]

So far in this book we've talked a lot about stars and planets forming. But, just as they are born, stars must also die. How long a star lives for depends on its mass. Higher-mass stars have more hydrogen to fuel their nuclear furnaces, but they burn through this fuel at a much faster rate than lower-mass stars. This means that the lower mass a star is, the longer it lives for.

What happens at the end of a star's life also depends on its mass. The highest-mass stars (more than eight times the mass of the Sun) end their lives as massive supernova explosions, leaving only a dense remnant in the middle. This can either be a neutron star – a dense object the breadth of a medium-sized city like Glasgow or Frankfurt but with more mass than the Sun – or even a small black hole. Stars like the Sun will end their lives in a way that is less dramatic, but which still produces awe-inspiring vistas.

Stars like the Sun live for billions of years, slowing burning hydrogen in their cores and turning it into helium. During this time they stay at roughly the same size, temperature and brightness. Once the star has burned nearly all the hydrogen in its core it starts to evolve. The hydrogen fusion stops, taking away the heat source that supported the star. The core now shrinks and becomes denser. The only thing left supporting the core is a weird effect caused by the behaviour of electrons. Earlier in this book we discussed how atoms were like stepped amphitheatres with electrons moving up and down between different levels of seating. Well, the amphitheatre has pretty limited capacity. The electrons in an atom cannot all sit down at the front to get close to the nucleus, only a certain number can sit at each level. The electrons may all want to sit at the front, but only two lucky ones get the chance. The core of an evolving star is a bit more complex than our amphitheatre model of a single atom. Here the electrons are not restricted to an atomic amphitheatre like in a gas in a star's atmosphere. Yet like in an atom there are a number of different states (like the levels in the amphitheatre) and only two electrons can fit in each of these.

In the core of our evolving star there are a limited number of spaces available for electrons. The electrons fill up the lowest energy states first, meaning some are forced to stay at higher energies. These higher energies mean the electrons have a higher momentum than if they had all been able to stay in the lowest energy state. Pressure is, on the microscopic level, particles bumping into things. The air pressure you feel right now comes from trillions upon trillions of tiny molecules colliding with your skin. The more momentum the particles have, the higher the pressure. In the core of our evolving star, the momentum of the electrons that have been forced to stay at higher energies produces a pressure known as degeneracy pressure. In most physical situations, degeneracy pressure is much smaller than the normal gas pressure you would get from molecules colliding with your skin. The core of an evolving star, however, is very dense and here the degeneracy pressure gets so big that it can slow the collapse of the star's core.

Just outside the core there is a thin layer of hydrogen left where it is hot enough for fusion to happen. Beyond this layer the rest of the star puffs up so it gets bigger. As it does so its surface temperature cools. The star has become what is known as a red giant.

The Sun has a radius of a little under 700,000 kilometres (435,000 mi.). That is about one two-hundredth of the distance from the Earth to the Sun. Once the Sun becomes a red giant in about 5 billion years, it will expand to the point where its atmosphere almost reaches the orbit of the Earth. This will swallow Venus and Mercury and, possibly, eventually the Earth.

A star does not just sit as a red giant, it continues to evolve. The core heats as it slowly collapses and eventually it reaches the temperature where the helium itself ignites. This helium is quickly burned into carbon and oxygen. Once the helium fuel runs out the core is left supported by degeneracy pressure.

In the outer layers of the star, shells burning both hydrogen and helium produce huge pulsations that travel to the star's outer layers. Here these pulsations catch dust particles that have formed in the star's cooler upper atmosphere like gusts of wind catching leaves. These gusts become so violent that the star throws off its outer layers. This produces spectacular ringed structures known as planetary nebulae. These have nothing to do with planets, but when they were first discovered they looked like how planets forming around a star were expected to look. The name stuck despite its inaccuracy.

Once a star has thrown off its outer layers the carbon and oxygen core in the middle remains. This is supported by degeneracy pressure and is known as a white dwarf. The white dwarf does not have an internal energy source from nuclear fusion, so it shines by radiating away the heat left over from its formation. It retains a large fraction of the mass from its parent star but, being only supported by degeneracy pressure, its atoms are squeezed together very tightly. White dwarfs are small compared to stars, about the size of the Earth. This means that they are very dense: a teaspoon of material from a white dwarf would have the same mass as a heavy goods vehicle.

White dwarfs are not usually only made of carbon and oxygen. There is normally a bit of hydrogen and helium left over of the stars they formed from. As white dwarfs are massive and small they have very strong gravity. This causes the heavier carbon and oxygen to settle in the centre, with helium and hydrogen forming the outer layers.

Astronomers looking at spectroscopic observations of white dwarfs divide them into types. Some show hydrogen in their atmospheres, some helium. But a few show something odd, metals and other heavier elements. That is weird, because these elements should sink quickly to the centre of the white dwarf, not hang around in their atmosphere leaving their mark on the white dwarf's spectrum.

It is possible a white dwarf could have been polluted by material it encountered as it moved through the Galaxy, but there are not enough metals floating about between the stars to explain all the polluted white dwarfs we see. As with so much of science, it is the phrase 'that is weird!' that is the portent for something very, very interesting indeed.

Now is a good time to introduce the particular white dwarf that is the focus of this chapter. GD 362 is about 160 light years from Earth. Like many other white dwarfs, it shows signs of metal pollution in its spectrum.

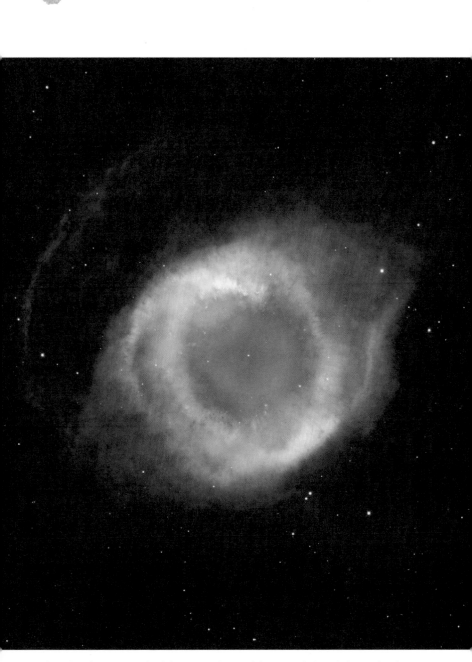

The Helix Nebula, an example of the erroneously named planetary nebulae, which are in fact the death throes of a star of similar mass to the Sun. After the outer layers have been thrown off only a tiny dead star called a white dwarf remains in the middle.

White dwarfs are generally quite hot (GD 362 has a temperature of about 10,000°C (18,032°F)) so give off most of their radiation in blue light. In 2005 astronomers looked at GD 362 and sure enough they found a lot of blue light.[2] However, they saw more infrared light than they were expecting. The best explanation for this excess of infrared light was a disc of glowing dust around GD 362 with a temperature of around 600°C (1,112°F).

Astronomers then took a spectrum of GD 362 in infrared light using the Spitzer Space Telescope. This showed signs of silicate material.[3] Silicates are one of the main components of the rocks that make up terrestrial planets and asteroids.

When astronomers looked in more detail at the fingerprints left by metals on the visible light spectrum of GD 362, they found what appeared to be rock in its atmosphere. It is sensible to assume that the rocky material from the disc is polluting the white dwarf with metals. So how did this disc of rocky material come to orbit the white dwarf?

The solar system may look like a clockwork toy with the planets going around the Sun in well-separated orbits. It is, however, a much more chaotic system than it might seem at first. The larger planets have such strong gravity that they kick around the smaller bodies like asteroids. In the asteroid belt between Mars and Jupiter the gravitational pull of both the Sun and Jupiter herds the asteroids into patterns of gaps and bands. At a distance of 2.3 Earth–Sun distances you will find lots of asteroids, but go to 2.5 Earth–Sun distances and you won't find many. This gap is the result of Jupiter's gravity making the orbit of anything going around the Sun at a distance of 2.5 Earth–Sun distances unstable. An asteroid that finds itself in this gap will be quickly kicked into another orbit by Jupiter's gravity.

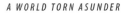

The last stage of the Sun's evolution towards becoming a white dwarf will be when it throws off its outer layers. This will produce a spectacular planetary nebula similar to the Helix Nebula, with around half of the Sun's mass being carried away. The white dwarf that remains will have much lower mass than the Sun does now. This means that its gravitational pull will be less. As a result, the orbits of the planets around the Sun will expand.

The positions of the gaps and bands in the asteroid belt between Mars and Jupiter are determined by the gravitational pull of the Sun and Jupiter. Once the Sun becomes a white dwarf its gravitational pull will be less, so the positions of these gaps and bands will change. Some asteroids will suddenly find themselves in a gap and thus in an unstable orbit. They might then be kicked into an elliptical orbit that takes them closer to the white dwarf that formed when the Sun died.

This is likely to be what happened to the planet that is the subject of this chapter, an asteroid that orbited the star that would become GD 362 for millions of years. Once that star died and became a white dwarf the asteroid was sent into an elliptical orbit. This brought it close enough to the white dwarf that it felt a much stronger gravitational pull on the side nearer the white dwarf than on the side facing away. The difference between these gravitational pulls is so big that it shredded the asteroid into dust and rubble. Now we see this former asteroid as a disc of material around a white dwarf and a few faint lines polluting its atmosphere.

THIS CHAPTER BEGAN with the story of an astronomer classifying a star in 1917. That particular star ended up bearing the astronomer's name (van Maanen's star) but his classification of it as an F-type was not quite right. Admittedly it had metal lines

An artist's impression of an asteroid torn to shreds by the strong gravity of a white dwarf similar to GD 362. The material from the asteroid forms a disc around the white dwarf and some of this debris will eventually pollute the dead star's atmosphere.

in its spectrum like an F-type star would be expected to have, but it was far, far too faint. In fact it turned out to be the first metal-polluted white dwarf, an object being showered with rocks from a shredded planet. This was probably the first observational evidence for an exoplanet system, but this was not realized for ninety years.

And speaking of firsts . . .

A DIAMOND BORN FROM DEATH

The worlds found around other stars are a curious collection. While some of them have been possibly like our own familiar Earth, others are truly bizarre, alien worlds. 51 Peg b, the first planet found around another star similar to the Sun, may have seemed alien enough, but three years before, worlds had been discovered around a star that is very different from the Sun. These planets are possibly still the most bizarre worlds we know of.

In the last chapter we met dying stars. Stars like the Sun will burn hydrogen for most of their lives and then slowly change into a red giant. In their final phase, before becoming a white dwarf, these stars have cores made of carbon and oxygen surrounded by layers of hydrogen and helium.

Stars more massive than about eight times the mass of the Sun die in a slightly different way. More massive stars do not stop their evolution with a core of carbon and oxygen, they keep fusing elements. First, they produce silicon, then nickel and iron. The star ends up with a structure like an onion, with the heaviest elements in the core surrounded by layers of progressively lighter elements.

The star has now driven itself into a dead end. Fusing iron into new elements will not release energy and hence will not provide the heat to prevent the star from collapsing. The star's core is so massive that even electron degeneracy pressure – that weird effect caused by the limited number of spaces for electrons at each energy level – cannot prevent it from collapsing. This collapse drives a shockwave that throws away the outer layers of the star in a supernova explosion. Left in the middle is either a black hole or a neutron star, an object a million times denser than a white dwarf. It is so dense that a 5-centimetre-thick layer of neutron star material covering the city of Amsterdam would have the same mass as the Earth.

In Chapter Ten we met a student spinning on a chair who twirled faster when they pulled the two textbooks they held at arm's length towards their body. The same principle applies to a neutron star. The core of a star prior to a supernova explosion would have been spinning. The collapse of the core pulled material towards its centre. Like the student pulling the textbooks towards their body, the collapse of the core speeds up its rotation, and I mean really speeds it up. While a star may have a rotation period of a few days, a newly formed neutron star will rotate a few times a second.

Neutron stars are pretty weird places. As the name suggests, they are formed mostly of particles called neutrons and are supported by another kind of degeneracy pressure, neutron degeneracy pressure. A neutron star is small, only about 10 kilometres (6 mi.) across but as massive as a star. This means that at its surface gravity is pretty strong, making its surface pretty smooth. A mountain on a neutron star would only get to be a few centimetres high before it was crushed under its own weight. Neutron stars also have strong magnetic fields. If the star from which the neutron star formed had a magnetic field this would stay with the neutron star after the supernova explosion. But the neutron star is much smaller than the star it formed from. Rather than being a giant star about 300 million kilometres (185 million mi.) across, it is a dense little ball the diameter of a medium-sized city. This means the star's magnetic field is squashed and squeezed until it wraps round the neutron star like a tense, taut, coiled spring.

The Earth also has a magnetic field, which is what makes a compass point north, at least provided you are not too far north. The Earth's magnetic field is not exactly aligned with its axis of rotation. While the north pole of the Earth's rotation is at, well, the North Pole, the north pole of its magnetic field is about two and a half degrees away in the direction of the Bering Strait.

The magnetic field of a neutron star might also not be perfectly aligned with its axis of rotation. This leads to an interaction between the strong magnetic field and the spinning neutron star, causing any charged particles in the neutron star to race towards the magnetic poles. There they thrash about in the magnetic field emitting radio waves. These travel away from the neutron star and out into the cosmos.

As the neutron star rotates an observer on the Earth will sometimes see a magnetic pole pointing towards them, but sometimes not. This results in flashes of radio waves being seen at regular intervals. This has led to neutron stars being known as pulsars or cosmic lighthouses, turning their lanterns towards the Earth many times a second.

Signals at regular intervals are the hallmark of a pulsar. By using what was at the time the world's largest radio telescope on Puerto Rico, astronomers noticed one pulsar known as PSR B1257+12 showed a strange pattern.[1] Sometimes its flashes arrived before they were expected, sometimes after. They would be more and more ahead of schedule for a few weeks, then less ahead of schedule, then behind schedule. So, what was moving around the timing of this pulsar's flashes?

The first planets around Sun-like stars were discovered with the radial velocity method. This is based on the fact that objects in a planetary system do not orbit the star in the middle but a balance point known as the centre of mass. This includes the star, moving in a tiny orbit. When viewed from the Earth the star would move away from us, then towards us and back again.

If there were a planet around PSR B1257+12 the pulsar would orbit the centre of mass of the pulsar-planet system, the pivot point similar to the one the rhino on the see-saw had earlier. First it would go towards an observer sitting on the Earth, then away from them. As it came closer each flash would have a shorter distance to travel to get to the Earth, so it would arrive ahead of schedule. As it moved away each pulse would start to arrive later and later. First ahead of schedule, then back to schedule, then behind schedule – exactly what is seen in the behaviour of PSR B1257+12.

By studying the changes in the timing of the flashes of PSR B1257+12, astronomers worked out in 1992 that this pulsar was orbited by two planets. These both had masses about four times the mass of the Earth and had orbital periods of 66 and 98 days. Note the discovery date: these were the first confirmed planets outside our solar system, found three years before the discovery of 51 Peg b. In 1994 further analysis of the data identified a third, even lower-mass planet, one with a mass a bit more than our own Moon and an orbital period of 25 days.[2] This world, PSR B1257+12 b (this chapter's planet), still holds the record as the lowest-mass planet discovered around another star.

So how did three planets come to be orbiting a pulsar. The obvious answer is that they were orbiting the star that ultimately blew up to form the pulsar. Unfortunately, the supernova explosion would have not only shattered the star, but should have simultaneously blasted away any surrounding planets. This puzzle left astronomers looking to fit together three elements of the mystery of this pulsar's planets.[3]

The Crab Nebula, bubbles and filaments of gas blasted out by an exploding star — also known as a supernova that was observed in the year 1054. At its heart lies a small spinning neutron star similar to PSR B1257+12.

First, if the planets were formed in orbit around the pulsar then you need to get enough material close to the pulsar to form planets.

Second, PSR B1257+12 is a special kind of pulsar. The process that fuels the radio emission from the pulsar's magnetic poles carries away energy. This slows the rotation and reduces the inter-action with the magnetic field, cutting the energy supply to the pulsar's poles. Eventually, after 10 million years or so the pulsar stops sending out radio flashes. This time frame is less than 1 per cent of the age of the Galaxy, meaning there is a vast graveyard of dead pulsars that no longer flash like cosmic lighthouses. It is, however, possible for a pulsar to rise from the grave. A pulsar with a star or white dwarf orbiting it can accrete matter from its companion. This is like our spinning student on a chair having textbooks lobbed at them. Every textbook they catch adds more rotational momentum, speeding up their rotation. That's what makes a pulsar rise from the grave, it gobbles matter from a com-panion. This adds rotational momentum and spins the pulsar up, bringing it back to life, spinning even faster than it did when it was first born. PSR B1257+12 is one such resurrected pulsar, so any model for planet formation needs to include this rise from the grave.

Finally, planets around pulsars are pretty rare. Perhaps less than 1 per cent of these strange stellar remnants have planets around them. Hence any model would need to explain why we do not see planets around every pulsar. There are some explan-ations that have been proposed that do not pass all three tests.

The pulsar could have passed very close to another star as it moved through the Galaxy. Such events are rare, so that's that box ticked. The planets around the star it encounters would be captured by the pulsar, which explains how the planets formed.

An artist's impression of the planets orbiting the pulsar PSR B1257+12. The bright cones of light pointing in opposite directions sweep around as the pulsar spins, sending regular flashes into space.

The problem comes with the next bit. The pulsar would sink into the middle of the star and start eating away at it from the inside. Eventually this would disrupt the star, leaving only a pulsar with planets around it. The pulsar would spin up a bit as it consumed the star from inside, but it is unlikely it would spin it up to the speed at which PSR B1257+12 is rotating.

Another possibility is that the material from the supernova explosion could fall back and form a disc around the pulsar. New planets could then be born in this disc. However, this process is unlikely to get enough matter back around the pulsar to form the planets we see around PSR B1257+12.

The best solution astronomers have come up with relates to one of the most significant discoveries of the last few years, gravitational waves. Any object in orbit around another object emits these. Gravitational waves carry away energy and make the objects spiral towards each other. The effect can be tiny: over the last 4 billion years or so the emission of gravitational waves has caused the Earth to move closer to the Sun by the gargantuan distance of 2 millimetres (0.08 in.). The more massive the objects involved are, the bigger the effect. Also, the closer the objects are to each other the bigger the effect.

The last few years have seen the detection of bursts of gravitational waves from neutron stars (or black holes) in orbit around each other. These pairs of astrophysical heavyweights lose so much energy to gravitational waves that they spiral together and collide.

It is possible that the star that went on to form PSR B1257+12 had another, lower-mass star in orbit around it. After a while this star would also die. As it is lower in mass it would end its life as a white dwarf. Over time the white dwarf and PSR B1257+12 would lose energy to gravitational waves and spiral towards each other. Rather than the two merging, once the white dwarf got very close to PSR B1257+12 it would feel a much bigger gravitational pull on the side closer to the neutron star than on the side facing away. This would shred the white dwarf into pieces, much like GD 362 was shredding asteroids in the last chapter.

The material from the disintegrated white dwarf would form a massive disc of material around PSR B1257+12. Some of this material would be dumped on PSR B1257+12, spinning it up and turning on the lantern of this cosmic lighthouse. Some could form planets in a similar way to the material in the disc around HL Tau. These planets would be made only from the material from the destroyed white dwarf. This would be mostly carbon and thus the planets will have crystallized carbon cores. This means the planets around PSR B1257+12 would be 100 thousand trillion trillion carat diamonds.[4]

This process can get enough material to the pulsar to both form the planets and to spin up the pulsar, making it rise from the grave. The relatively small number of pulsars with a close white dwarf companion also means this process passes the rarity test.

The first confirmed planets outside our solar system are possibly still the weirdest worlds we have encountered: massive diamonds orbiting a zombie pulsar. The future may hold planets that are more familiar, or perhaps more bizarre.

EPILOGUE:
YET MORE WORLDS

We have met twenty extraordinary worlds – from the first bizarre worlds discovered around other stars to planets more like our own Earth. We have seen hidden worlds forming, worlds that could host life and worlds dying.

The telescopes that discovered these planets have ranged from telephoto lenses to one of the Earth's largest radio telescopes. There have been worlds discovered with purpose-built space missions and worlds discovered with old telescopes repurposed for the study of exoplanets.

The next decade will see several large programmes to both discover and characterize exoplanets. Some of these use dedicated space missions built to match those goals, some will use advanced instruments on new general-purpose telescopes.

The Transiting Exoplanet Survey Satellite (TESS) launched in April 2018. This NASA mission is scanning right across the sky for transiting exoplanets around the brightest stars. The Kepler mission discovered many exoplanets around relatively faint stars. The faintness of these parent stars made it difficult or impossible to determine the masses of these planets with radial velocity measurements or to characterize their atmospheres with transit spectroscopy. TESS is designed to discover hundreds of potentially rocky planets around bright stars. These planets will be easier to characterize than those found by Kepler around fainter stars. In addition, TESS will discover thousands more planets, both larger gas or ice giants around bright stars and planets around fainter stars. TESS may find around ten potentially rocky worlds in the habitable zones around their stars. Surveying the sky bit by bit, TESS will look for planets in one part of the heavens and will then move on. Most of the sky will only be observed for a continuous period of four weeks. Thus TESS will mostly discover planets in short orbits. This means all the habitable zone planets it will discover will be around small red stars similar to Proxima Centauri or TRAPPIST-1 as these stars have their habitable zones closer to them.[1] As of 31 January 2020 TESS has surveyed over three-quarters of the sky and discovered 38 planets with 1,660 planet candidates awaiting confirmation.[2] These include three small, likely rocky worlds orbiting a small red star only 35 light years from the Earth;[3] a small, likely rocky world in a multiple star system consisting of three small red stars around 23 light years from Earth;[4] and a gas giant planet orbiting a star in a young moving group that is only 45 million years old.[5]

The other specially designed exoplanet discovery mission is PLATO (PLAnetary Transits and Oscillations of Stars). This is a European Space Agency mission that will launch in the mid-2020s. PLATO will consist of an array of 26 small telescopes in one spacecraft.[6] These will allow PLATO to stare at a much larger patch of sky than Kepler. Each PLATO observation will cover 5 per cent of the sky; Kepler's field of view was one-twentieth of that area. This wider field will allow PLATO to observe a far higher number of bright stars. Any planets discovered around these bright stars will, like the TESS bright star discoveries, be much easier to characterize than the typical Kepler discovery. PLATO will also stare at the same patch of sky for longer than TESS will. This means it will be able to discover planets with much longer orbital periods than TESS. Hence it will be able to discover rocky planets in the habitable zones of Sun-like stars. These worlds will be less affected by hostile stellar environments than a planet like Proxima b. Over time PLATO will change the field it observes. Depending on the observing strategy that is chosen, it will survey between about one-tenth and roughly one-half of the sky during its four-year mission.

Both the PLATO and TESS missions will discover many more worlds. The pair of them have the potential to increase the number of rocky worlds we know about in the habitable zones of other stars. This means astronomers will be able to measure the occurrence rate of planets similar to the Earth more accurately. They will tell us how many worlds of comparable size to ours there are in the Galaxy. They will not, however, tell us if these planets have hospitable atmospheres.

NEWLY FOUND worlds need to be characterized. What mass is this planet? Is it rocky? Does it have an atmosphere?

There are currently two planned or recently launched space missions to characterize exoplanets. One is CHEOPS (CHaracterising ExOPlanet Satellite), a recently launched Swiss-led European mission to get extremely precise radius measurements of transiting exoplanets by determining their transit depths with great accuracy. These measurements will improve on the estimates of planetary size achievable from the ground. The other is ARIEL (Atmospheric Remote-sensing Infrared Exoplanet Large-survey), a European Space Agency mission due to launch in 2028. This is a transit spectroscopy mission that will study the atmospheres of planets found by the likes of PLATO and TESS.

The James Webb Space Telescope (JWST) will be launched in 2021. JWST is a NASA mission with contributions from Europe and Canada. It is a multi-purpose observatory that can study anything from planets in our own solar system to the most distant galaxies. Its instruments will allow astronomers to study planetary atmospheres by transit spectroscopy.

JWST will also be able to directly image planets around other stars. Being outside the atmosphere will mean that observations with JWST will not require the myriad of complex corrections for atmospheric blurring that have to be done from the ground. JWST will simply use a blocking circle known as a coronagraph to mask out the light from a planet's parent star. The worlds JWST directly observes will not be rocky Earth-like worlds, but it is capable of imaging lower-mass worlds than the huge gas giants to which ground-based observations are currently limited.

The proposed NASA WFIRST (Wide Field Infrared Survey Telescope) mission is also planned to have a coronagraph, hence it will also be able to directly image planets. There are also plans for WFIRST to undertake a microlensing survey, allowing it to detect planets from the brief flashes they cause as their gravity bends light from a background star. Microlensing is much more sensitive to small planets in relatively wide orbits than other planet-finding techniques.

One other general-purpose space observatory that will help to discover planets has already been launched. The European Gaia satellite is accurately measuring the positions of billions of stars. This will allow it to detect the planets around the nearest stars in a way we have not met yet. The radial velocity method discovers planets by observing the movements towards and away from the Earth of the tiny orbit a star is pulled in by its planet. Gaia will measure the positions of stars in a way that is accurate enough to detect the rest of that motion, the side-to-side movement. This will be especially good for detecting big planets in fairly wide orbits.

Finally, there is big glass on the ground. Telescope technology has advanced to the point of building huge mirrors, tens of metres across. Three large telescopes are currently either proposed or under construction. The European Southern Observatory's Extremely Large Telescope and the Giant Magellan Telescope will study the southern hemisphere and the Thirty Metre Telescope will cover the northern hemisphere.

These telescopes will come equipped with new, more accurate instruments and better adaptive optics systems. This means they will be able to measure radial velocities more accurately and hence both discover smaller planets and get superb mass measurements for the worlds discovered by TESS and PLATO. The new adaptive optics systems will improve the corrections for atmospheric blurring, allowing these telescopes to directly image planets closer to their parent stars and to get pictures of smaller, fainter planets.

In combination these space missions and large ground-based telescopes will discover more worlds and inform us more about them. We will get a better estimate of how many rocky worlds lie in the habitable zones around different types of stars. We will also get a better idea of what other planetary systems look like. Many systems discovered by the transit method contain only one planet. Do these stars have only one planet or do they have other worlds in wider orbits we haven't found yet? Will we ever find an analogue of our own solar system with multiple terrestrial planets, gas giants and ice giants?

We will also learn more about the individual planets themselves. We will find out more about the atmospheres of hot Jupiters, if rocky planets have atmospheres similar to Earth, maybe even if planets show the telltale signs of chemicals produced by life.

SO THAT'S IT, in less than thirty years astronomers have gone from knowing no other worlds around other stars to knowing thousands. The search for planets around other stars has gone from a relatively niche activity to being one of the most dynamic and active areas in all of science.

One thing is for sure, the future I have laid out here is an incomplete one. This is for two reasons. First, science is a creative process and astronomers will always think of clever new ways to use telescopes both old and new. Second, the universe always surprises us. We have not yet seen a set of carbon copies of the solar system around other stars. We have seen a huge range of weird and wonderful worlds, greater than we could have imagined a little more than a quarter of a century ago. Since the discovery of the first planets around other stars we have had our expectations challenged and proved wrong, again and again. The future will bring more invalidated assumptions and destroyed preconceptions.

There are more worlds out there, more diverse and wonderful than we can currently comprehend.

APPENDIX: THE TWENTY WORLDS

BELOW IS a table of the twenty worlds that are the subject of this book. Also presented are planets in the same systems as the twenty worlds (shown in italics). The Sun and solar system planets are shown for comparison. The masses and radii of the planets are scaled so that the Earth has a value of one for both parameters. The distance from the parent star quoted is in Astronomical Units (AU). The Earth's typical distance from the Sun (in technical terms the semi-major axis of its orbit) is 1AU. Some of the parameters listed are measurements with high uncertainties, some are estimates based on a number of assumptions about the planet's parent star. However they give a good overview of the type of orbit each planet is in and what its characteristics are physically.

Name	Period (days)	Distance from parent star (AU)	Mass (Earth masses)	Radius (Earth radii)	Composition or type	Parent star	Parent star type	Date discovered
Sun	–	–	333,000	109	Star	–	–	–
Mercury	88.0	0.387	0.06	0.38	Rocky	Sun	Yellow dwarf	–
Venus	225	0.723	0.82	0.95	Rocky	"	"	–
Earth	365	1.00	1.00	1.00	Rocky	"	"	–
Mars	687	1.52	0.11	0.53	Rocky	"	"	–
Jupiter	4,333	5.20	318	11.2	Gas giant	"	"	–
Saturn	10,759	9.58	95.2	9.45	Gas giant	"	"	–
Uranus	30,685	19.2	14.5	4.01	Ice giant	"	"	1781
Neptune	60,189	30.0	17.2	3.88	Ice giant	"	"	1846
ALIEN WORLDS								
51 Peg b	4.23	0.052	>146	21.3	Gas giant	51 Peg	Yellow dwarf	1995[1]
HD 209458 b	3.52	0.047	219	15.5	Gas giant	HD 209458	Yellow dwarf	2000[2]

Name	Period (days)	Distance from parent star (AU)	Mass (Earth masses)	Radius (Earth radii)	Composition or type	Parent star	Parent star type	Date discovered
HD 189733b	2.22	0.031	363	12.7	Gas giant	HD 189733	Orange dwarf	2005[3]
WASP-19b	0.79	0.016	354	15.6	Gas giant	WASP-19	Yellow dwarf	2010[4]
HAT-P-7b	2.20	0.038	554	16.0	Gas giant	HAT-P-7	Whitish yellow dwarf	2008[5]
TOWARDS EARTH								
OGLE-2005-390L b	3,500	2.1	5.41	-	Super-Earth	OGLE-2005-390L	Red dwarf	2006[6]
Kepler-9b	19.2	0.14	43.5	8.20	Gas giant	Kepler 9	Yellow dwarf	2010[7]
Kepler-9c	39.0	0.23	29.9	8.29	Gas giant	"	"	"
Kepler-9d	1.590	0.027	5.25	2.00	Possible Super-Earth	"	"	"
Kepler-36b	13.8	0.12	4.46	1.479	Super-Earth, possibly rocky	Kepler 36	Yellow subgiant	2012[8]
Kepler-36c	16.2	0.13	8.08	3.68	Ice giant	"	"	"
Kepler-10b	0.837	0.017	3.33	1.468	Rocky	Kepler-10	Yellow dwarf	2011[9]
Kepler-10c	45.3	0.24	7.37[10]	2.349	Possible water world	"	"	2011[11]
BIRTH								
Probable planetary embryo around HL Tau	–	–	–	–	Likely a gas giant in the process of formation	HL Tau	Very young higher-mass star	–
beta Pic b	8,200	9.7	4,040	18.2	Gas giant	beta Pictoris	Young higher-mass star	2009[12]
HR 8799 b	164,000	68	2,226	13.4	Gas giant	HR 8799	Young higher-mass star	2008[13]
HR 8799 c	82,100	43	2,639	14.6	Gas giant	"	"	"
HR 8799 d	41,100	27	2,639	13.4	Gas giant	"	"	"
HR 8799 e	18,000	16	2,925	13.1	Gas giant	"	"	"
PSO J318.5-22	–	–	2,703[14]	–	Gas giant	–	–	2013[15]

Name	Period (days)	Distance from parent star (AU)	Mass (Earth masses)	Radius (Earth radii)	Composition or type	Parent star	Parent star type	Date discovered
WISE 0855-0714	–	–	1908	–	Gas giant	–	–	2014[16]
LIFE								
TRAPPIST-1b	1.511	0.012	1.017	1.127	Probably rocky, possible water envelope	TRAPPIST-1	Red dwarf	2016[17]
TRAPPIST-1c	2.422	0.016	1.156	1.100	Rocky	"	"	"
TRAPPIST-1d	4.05	0.022	0.297	0.788	Not well constrained but likely to be rocky	"	"	"
TRAPPIST-1e	6.100	0.029	0.772	0.915	"	"	"	2017[18]
TRAPPIST-1f	9.206	0.038	0.934	1.052	Likely rocky with possible ice sheet	"	"	"
TRAPPIST-1g	12.35	0.047	1.148	1.154	"	"	"	"
TRAPPIST-1h	18.77	0.062	0.331	0.777	"	"	"	"
Proxima b	11.19	0.049	>1.3	–	Possibly rocky	Proxima Centauri	Red dwarf	2016[19]
DEATH								
WASP-12b	1.091	0.023	467	21.28	Gas giant	WASP-12	Yellow dwarf	2009[20]
Hypothetical asteroid disrupted by GD 362	–	–	–	–	Disrupted rocky asteroid	GD 362	White dwarf	–
PSR B1257+12 b	25.3	0.19	0.02	–	Likely crystallized carbon	PSR B1257+12	Pulsar	1994[21]
PSR B1257+12 c	66.5	0.36	4.134	–	"	"	"	1992[22]
PSR B1257+12 d	98.2	0.46	3.816	–	"	"	"	"

NOTES: All solar system data comes from the NASA Goddard Lunar and Planetary Science website.[23] Exoplanet data is taken from the *Extrasolar Planet Encyclopedia* unless otherwise stated,[24] with the exception of the TRAPPIST-1 planets where they come from two other sources.[25] The quoted discovery date is the date of the published paper. Note that in the case of HD 209458 b this means the discovery date is one year after the date of the first transit observations. The stellar type quoted is a description of a star's colour (and hence temperature) and size. The composition or type is a description of the current best estimate for the likely composition of any particular object. No discovery dates are given for the hypothetical HL Tau planet and GD 362 disrupted asteroid.

REFERENCES

For further reading on the topic, see Michael Perryman's *The Exoplanet Handbook*, 2nd edn (Cambridge, 2018) and Elizabeth Tasker's *The Planet Factory* (London, 2019).

INTRODUCTION: AN ORDERED FAMILY PORTRAIT

1 S. B. Gaudi et al., 'A Giant Planet Undergoing Extreme-ultraviolet Irradiation by its Hot Massive-star Host', *Nature*, 546 (2017), pp. 514–18.
2 Wayne Horowitz and Alexandra Horowitz, *Mesopotamian Cosmic Geography* (University Park, PA, 1998), p. 153.
3 'Planet', *Lexico*, www.lexico.com, accessed 30 September 2019.
4 Louis Strous, 'Who Discovered that the Sun was a Star?', Stanford Solar Center, http://solar-center.stanford.edu, accessed 29 March 2019.
5 Joseph Needham and Wang Ling, *Science and Civilisation in China*, vol. III: *Mathematics and the Sciences of the Heavens and Earth* (Cambridge, 1959), p. 227.
6 Jonathan J. Fortney, 'Looking into the Giant Planets', *Science*, 305 (2004), pp. 1414–15.
7 Marius Millot et al., 'Experimental Evidence for Superionic Water Ice Using Shock Compression', *Nature Physics*, XIV (2018), pp. 297–302.
8 '"The Taking of Christ" by Michelangelo Merisi da Caravaggio', www.nationalgallery.ie, accessed 29 March 2019.
9 'A 19th-century Vision of the Year 2000', https://publicdomainreview.org, accessed 29 March 2019.

ALIEN WORLDS

1 A WORLD BEYOND EXPECTATION

1 Michel Mayor and Didier Queloz, 'A Jupiter-mass Companion to a Solar-type Star', *Nature*, 378 (1995), pp. 355–9.
2 A. Baranne et al., 'ELODIE: A Spectrograph for Accurate Radial Velocity Measurements', *Astronomy and Astrophysics Supplement*, 119 (1996), pp. 373–90.

2 A WORLD OBSCURING

1 J. Wang and E. B. Ford, 'On the Eccentricity Distribution of Short-period Single-planet Systems', *Monthly Notices of the Royal Astronomical Society*, CDXVIII/3 (2011), pp. 1822–33.
2 D. Charbonneau et al., 'Detection of Planetary Transits Across a Sun-like Star', *Astrophysical Journal*, DXXIX/1 (2000), pp. L45–L48. There is also an independent discovery of the transit in G. W. Henry et al., 'A Transiting "51 Peg-like" Planet', *Astrophysical Journal*, DXXIX/1 (2000), pp. L41–L44.
3 J. Southworth, 'Homogeneous Studies of Transiting Extrasolar Planets, III: Additional Planets and Stellar Models', *Monthly Notices of the Royal Astronomical Society*, CDVIII/3 (2010), pp. 1689–713.
4 K. Batygin and D. J. Stevenson, 'Inflating Hot Jupiters with Ohmic Dissipation', *Astrophysical Journal Letters*, DCCXIV/2 (2010), pp. L238–L243.

3 A TEMPESTUOUS WORLD

1 François Bouchy et al., 'ELODIE Metallicity-biased Search for Transiting Hot Jupiters, II: A Very Hot Jupiter Transiting the Bright K Star HD 189733', *Astronomy and Astrophysics*, CDXLIV/1 (2005), pp. L15–L19.
2 Heather A. Knutson et al., 'A Map of the Day–Night Contrast of the Extrasolar Planet HD 189733b', *Nature*, 447 (2007), pp. 183–6.
3 Heather A. Knutson et al., 'Multiwavelength Constraints on the Day–Night Circulation Patterns of HD 189733b', *Astrophysical Journal*, DCXC/1 (2009), pp. 822–36.
4 Adam P. Showman and Lorenzo M. Polvani, 'Equatorial Superrotation on Tidally Locked Exoplanets', *Astrophysical Journal*, DCCXXXVIII/1 (2011), id. 71.

4 A GLIMMER OF ATMOSPHERE

1 L. Hebb et al., 'WASP-19b: The Shortest Period Transiting Exoplanet Yet Discovered', *Astrophysical Journal*, DCCVIII/1 (2010), pp. 224–31.
2 L. Mancini et al., 'Physical Properties, Transmission and Emission Spectra of the WASP-19 Planetary System from Multi-colour Photometry', *Monthly Notices of the Royal Astronomical Society*, CDXXXVI/1 (2013), pp. 2–18.
3 Paul Sutherland, 'Three New Planets, Thanks to eBay', www.skymania.com, 1 November 2007.
4 A. M. Mandell et al., 'Exoplanet Transit Spectroscopy Using WFC3: WASP-12 b, WASP-17 b, and WASP-19 b', *Astrophysical Journal*, DCCLXXIX/2 (2013), id. 128.
5 E. Sedaghati et al., 'Detection of Titanium Oxide in the Atmosphere of a Hot Jupiter', *Nature*, 549 (2017), pp. 238–41.
6 N. Espinoza et al., 'ACCESS: a Featureless Optical Transmission Spectrum for WASP-19b from Magellan/IMACS', *Monthly Notices of the Royal Astronomical Society*, CDLXXXII/2 (2019), pp. 2065–87.

5 A WORLD IN REVERSE

1 W. Somerset Maugham, *The Summing Up* (London, 1938), p. 235.
2 Stefan Coerts, 'Iniesta: There is No Such Thing as the Perfect Player', www.goal.com, 21 December 2012.
3 H. Kragh, 'The Source of Solar Energy, ca. 1840–1910: From Meteoric Hypothesis to Radioactive Speculations', *European Physical Journal H*, XLI/4 (2016), id. 394.
4 A. Pál et al., 'HAT-P-7b: An Extremely Hot Massive Planet Transiting a Bright Star in the Kepler Field', *Astrophysical Journal*, 680 (2008), pp. 1450–56.
5 O. Benomar et al., 'Determination of Three-dimensional Spin-orbit Angle with Joint Analysis of Asteroseismology, Transit Lightcurve, and the Rossiter-McLaughlin Effect: Cases of HAT-P-7 and Kepler-25', *Publications of the Astronomical Society of Japan*, LXVI/5 (2014), id. 9421.
6 J. N. Winn et al., 'Hot Stars with Hot Jupiters Have High Obliquities', *Astrophysical Journal Letters*, DCCXVIII/2 (2010), pp. L145–L149; A.H.M.J. Triaud,'The Rossiter-McLaughlin Effect in Exoplanet Research', in *Handbook of Exoplanets*, ed. Hans J. Deeg and Juan Antonio Belmonte (Cham, 2018), id. 2.

7 R. I. Dawson, 'On the Tidal Origin of Hot Jupiter Stellar
   Obliquity Trends', *Astrophysical Journal Letters*, DCCXC/2 (2014),
   id. L31.

TOWARDS EARTH

6 A FLASH FROM DARKNESS

1 Judith Ann Schiff, 'The Frisbee Files', *Yale Alumni Magazine*
   (May/June 2007).
2 Malcolm Longair, 'Bending Space–Time: A Commentary on
   Dyson, Eddington and Davidson (1920) "A Determination
   of the Deflection of Light by the Sun's Gravitational Field"',
   Philosophical Transactions of the Royal Society A, CCCLXXIII/2039
   (2015).
3 J.-P. Beaulieu et al., 'Discovery of a Cool Planet of 5.5 Earth
   Masses through Gravitational Microlensing', *Nature*, 439 (2006),
   pp. 437–40.

7 A WORLD YOU CAN'T SET YOUR WATCH BY

1 Ulinka Rublack, 'The Astronomer and the Witch – How Kepler
   Saved his Mother from the Stake', www.cam.ac.uk, 22 October
   2015.
2 Simon Winder, *Danubia: A Personal History of Habsburg Europe*
   (New York, 2014), p. 129.
3 M. J. Holman et al., 'Kepler-9: A System of Multiple Planets
   Transiting a Sun-like Star, Confirmed by Timing Variations',
   Science, 330 (2010), p. 51.
4 For an excellent analysis rejecting various blend scenarios
   for Kepler-9b, see G. Torres et al., 'Modeling Kepler Transit
   Light Curves as False Positives: Rejection of Blend Scenarios
   for Kepler-9, and Validation of Kepler-9 d, A Super-earth-size
   Planet in a Multiple System', *Astrophysical Journal*, 727 (2011),
   id. 24.

8 A CONTRASTING SIBLING

1 J. A. Carter et al., 'Kepler-36: A Pair of Planets with Neighboring
   Orbits and Dissimilar Densities', *Science*, 327 (2012), p. 556.
2 Ibid.

3 Ibid.

4 E. D. Lopez and J. J. Fortney, 'The Role of Core Mass in
   Controlling Evaporation: The Kepler Radius Distribution and
   the Kepler-36 Density Dichotomy', *Astrophysical Journal*, 776
   (2013), id. 2.

5 B. J. Fulton et al., 'The California Kepler Survey. III. A Gap in
   the Radius Distribution of Small Planets', *Astronomical Journal*,
   CLIV/3 (2017), id. 109.

9 A WORLD LIKE OURS?

1 N. M. Batalha et al., 'KEPLER's First Rocky Planet: Kepler-10b',
   Astrophysical Journal, DCCXXIX/1 (2011), id. 27.

2 X. Dumusque et al., 'The Kepler-10 Planetary System Revisited
   by HARPS-N: A Hot Rocky World and a Solid Neptune-mass
   Planet', *Astrophysical Journal*, DCCLXXXIX/2 (2014), id. 154.

3 The Research Consortium On Nearby Stars, 'RECONS Census
   of Objects Nearer than 10 Parsecs', www.recons.org, accessed
   19 August 2019.

4 E. A. Petigura et al., 'Prevalence of Earth-size Planets Orbiting
   Sun-like Stars', *Publications of the National Academy of Sciences*,
   LX/48 (2013), pp. 19273–8.

5 D. Foreman-Mackey et al., 'Exoplanet Population Inference
   and the Abundance of Earth Analogs from Noisy, Incomplete
   Catalogs', *Astrophysical Journal*, DCCXCV/1 (2014), id. 64.

6 C. D. Dressing and D. Charbonneau, 'The Occurrence of
   Potentially Habitable Planets Orbiting M Dwarfs Estimated from
   the Full Kepler Dataset and an Empirical Measurement of the
   Detection Sensitivity', *Astrophysical Journal*, DCCCVII/1 (2015),
   id. 45.

7 L. M. Weiss et al., 'The California-Kepler Survey, V: Peas in a
   Pod: Planets in a Kepler Multi-planet System Are Similar in Size
   and Regularly Spaced', *Astronomical Journal*, CLV/1 (2018), id.
   48.

8 J. Wang and D. Fischer, 'Revealing a Universal Planet-metallicity
   Correlation for Planets of Different Sizes Around Solar-type
   Stars', *Astronomical Journal*, CXLIX/1 (2015), id. 14.

9 A. L. Kraus et al., 'The Impact of Stellar Multiplicity on Planetary
   Systems, I: The Ruinous Influence of Close Binary Companions',
   Astronomical Journal, CLII/1 (2016), id. 8.

BIRTH

10 AN UNSEEN EMBRYO

1 The ALMA Partnership, 'The 2014 ALMA Long Baseline
   Campaign: First Results from High Angular Resolution
   Observations toward the HL Tau Region', *Astrophysical Journal
   Letters*, DCCCVIII/1 (2015), id. L3.

11 A WORLD THROUGH THE BLUR

1 A.-M. Lagrange et al., 'A Probable Giant Planet Imaged
   in the β Pictoris Disk. VLT/NaCo deep L'-band imaging',
   Astronomy and Astrophysics, CDXCIII/2 (2009),
   pp. L21–L25.
2 C.P.M. Bell et al., 'A Self-consistent, Absolute Isochronal
   Age Scale for Young Moving Groups in the Solar
   Neighbourhood', *Monthly Notices of the Royal
   Astronomical Society*, CDLIV/1 (2015), pp. 593–614.
3 A.-M. Lagrange et al., 'A Giant Planet Imaged in the Disk
   of the Young Star β Pictoris', *Science*, 329 (2010), p. 57.
4 M. Bonnefoy et al., 'Physical and Orbital Properties of
   β Pictoris b', *Astronomy and Astrophysics*, 567 (2014), id. L9.

12 AN EMBER SPAT FROM A FIRE

1 G. Chauvin et al., 'Giant Planet Companion to
   2MASSW J1207334-393254', *Astronomy and Astrophysics*,
   CDXXXVIII/2 (2005), pp. L25–L28.
2 P. Kalas et al., 'Optical Images of an Exosolar Planet 25
   Light-years from Earth', *Science*, 322 (2008), pp. 1345–8.
3 C. Marois et al., 'Direct Imaging of Multiple Planets Orbiting
   the Star HR 8799', *Science*, 322 (2008), pp. 1348–52.
4 C. Marois et al., 'Images of a Fourth Planet Orbiting HR 8799',
   Nature, 468 (2010), pp. 1080–83.
5 B. P. Bowler et al., 'Near-infrared Spectroscopy of the Extrasolar
   Planet HR 8799 b', *Astrophysical Journal*, DCCXXIII/1 (2010),
   pp. 850–68.
6 B. Zuckerman et al., 'The Tucana/Horologium, Columba,
   AB Doradus, and Argus Associations: New Members and
   Dusty Debris Disks', *Astrophysical Journal*, DCCXXXII/2
   (2011), id. 61.

7 C.P.M. Bell et al., 'A Self-consistent, Absolute Isochronal Age
Scale', *Monthly Notices of the Royal Astronomical Society*, CDLIV/1
(2015), pp. 593–614.

8 Ibid.; A. S. Binks and R. D. Jefferies, 'A Lithium Depletion
Boundary Age of 21 Myr for the Beta Pictoris Moving Group',
Monthly Notices of the Royal Astronomical Society, CDXXXVIII/1
(2014), pp. L11–L15.

9 S. E. Dodson-Robinson et al., 'The Formation Mechanism of Gas
Giant Planets on Wide Orbits', *Astrophysical Journal*, DCCVII/1
(2009), pp. 79–88.

13 A LONELY PLANET

1 M. C. Liu et al., 'The Extremely Red, Young L Dwarf
PSO J318.5338-22.8603: A Free-floating Planetary-mass Analog
to Directly Imaged Young Gas-giant Planets', *Astrophysical
Journal Letters*, 777 (2013), id. L20.

2 This is larger than the six Jupiter masses reported ibid., but takes
into account the increased age reported in A. S. Binks and
R. D. Jefferies, 'A Lithium Depletion Boundary Age of 21 Myr
for the Beta Pictoris Moving Group', *Monthly Notices of the Royal
Astronomical Society*, CDXXXVIII/1 (2014), pp. L11–L15.

3 V. Joergens et al., 'OTS 44: Disk and Accretion at the Planetary
Border', *Astronomy and Astrophysics*, 558 (2013), id. L7.

4 B. A. Biller et al., 'Variability in a Young, L/T Transition
Planetary-mass Object', *Astrophysical Journal*, 813 (2015), id. L23.

14 A DREICH WORLD?

1 'Dreich', *Lexico*, www.lexico.com, accessed 28 March 2019.

2 K. L. Luhman, 'Discovery of a ~250 K Brown Dwarf at 2 pc
from the Sun', *Astrophysical Journal*, 786 (2014), id. L18.

3 M. R. Zapatero Osorio et al., 'Near-infrared Photometry of WISE
J085510.74-071442.5', *Astronomy and Astrophysics*, 592 (2016),
id. A80.

4 C. V. Morley et al., 'An L Band Spectrum of the Coldest Brown
Dwarf', *Astrophysical Journal*, 858 (2018), id. 97.

5 K. L. Luhman and T. L. Esplin, 'The Spectral Energy
Distribution of the Coldest Known Brown Dwarf', *Astronomical
Journal*, 152 (2016), id. 78.

LIFE

15 A CURSED WORLD

1 M. Gillon et al., 'Temperate Earth-sized Planets Transiting a Nearby Ultracool Dwarf Star', *Nature*, 533 (2016), pp. 221–4.

2 L. Delrez et al., 'Early 2017 Observations of TRAPPIST-1 with Spitzer', *Monthly Notices of the Royal Astronomical Society*, CDLXXV/3 (2018), pp. 3577–97.

3 S. L. Grimm et al., 'The Nature of the TRAPPIST-1 Exoplanets', *Astronomy and Astrophysics*, 613 (2018), id. A68.

4 R. Luger et al., 'A Seven-planet Resonant Chain in TRAPPIST-1', *Nature Astronomy*, I (2017), id. 0129.

5 E. T. Wolf, 'Assessing the Habitability of the TRAPPIST-1 System Using a 3D Climate Model', *Astrophysical Journal*, DCCCXXXIX/1 (2017), id. L1.

16 A WORLD THAT'S JUST RIGHT?

1 M. Gillon et al., 'Seven Temperate Terrestrial Planets around the Nearby Ultracool Dwarf Star TRAPPIST-1', *Nature*, 542 (2017), pp. 456–60.

2 L. Delrez et al., 'Early 2017 Observations of trappist-1 with Spitzer', *Monthly Notices of the Royal Astronomical Society*, CDLXXV/3 (2018), pp. 3577–97.

3 S. L. Grimm et al., 'The Nature of the trappist-1 Exoplanets', *Astronomy and Astrophysics*, 613 (2018), id. A68.

4 'Assessing the Habitability of the trappist-1 System Using a 3D Climate Model', *Astrophysical Journal*, DCCCXXXIX/1 (2017), id. L1.

5 J. de Wit et al., 'Atmospheric Reconnaissance of the Habitable-zone Earth-sized Planets Orbiting TRAPPIST-1', *Nature Astronomy*, II (2018), pp. 214–19.

6 L. Kaltenegger, 'How to Characterize Habitable Worlds and Signs of Life', *Annual Review of Astronomy and Astrophysics*, LV (2017), pp. 433–85.

7 B. V. Rackham et al., 'The Transit Light Source Effect: False Spectral Features and Incorrect Densities for M-dwarf Transiting Planets', *Astrophysical Journal*, 853 (2018), id. 122.

8 C. T. Unterborn et al., 'Inward Migration of the TRAPPIST-1 Planets as Inferred from their Water-rich Compositions', *Nature Astronomy*, II (2018), pp. 297–302.

9 A. J. Burgasser and E. E. Mamajek, 'On the Age of the
TRAPPIST-1 System', *Astrophysical Journal*, 845 (2017), id. 110.

17 A WORLD BOMBARDED

1 G. Anglada-Escudé et al., 'A Terrestrial Planet Candidate in a
Temperate Orbit around Proxima Centauri', *Nature*, 536 (2016),
pp. 437–40.
2 T. E. Bell and T. Phillips, 'A Super Solar Flare', NASA Science
News, www.science.nasa.gov, 6 May 2008.
3 W. S. Howard et al., 'The First Naked-eye Superflare Detected
from Proxima Centauri', *Astrophysical Journal*, DCCCLX/2 (2018),
id. L30.
4 C. Garraffo et al., 'The Space Weather of Proxima Centauri b',
Astrophysical Journal, DCCCXXIII/1 (2016), id. L4.
5 I. Ribas et al., 'The Habitability of Proxima Centauri b, I:
Irradiation, Rotation and Volatile Inventory from Formation
to the Present', *Astronomy and Astrophysics*, 596 (2016), id. A111.
6 A. Segura et al., 'The Effect of a Strong Stellar Flare on the
Atmospheric Chemistry of an Earth-like Planet Orbiting an
M Dwarf', *Astrobiology*, X/7 (2010), pp. 751–71.
7 Ribas et al., 'Habitability of Proxima Centauri b'.
8 Ibid.

DEATH

18 A DARK CLOAK OF DEATH

1 L. Hebb et al., 'WASP-12b: The Hottest Transiting Extrasolar
Planet Yet Discovered', *Astrophysical Journal*, CDXCIII/2 (2009),
pp. 1920–28.
2 Calculated using the transit depths from C. A. Haswell et al.,
'Near-ultraviolet Absorption, Chromospheric Activity, and
Star-Planet Interactions in the WASP-12 system', *Astrophysical
Journal*, DCCLX/1 (2012), id. 79.
3 D. Locci, 'Photo-evaporation of Close-in Gas Giants
Orbiting around G and M Stars', *Astronomy and Astrophysics*,
624 (2019), A101.
4 S. W. Yee et al., 'The Orbit of WASP-12b is Decaying',
article-id:1911.09131, arxiv.org, accessed 4 December 2019.

19 A WORLD TORN ASUNDER

1 Elizabeth Landau, 'Overlooked Treasure: The First Evidence
of Exoplanets', www.jpl.nasa.gov, 1 November 2017.
2 M. Kilic et al., 'Debris Disks around White Dwarfs: The DAZ
Connection', *Astrophysical Journal*, DCXLVI/1 (2006), pp. 474–9.
3 M. Jura et al., 'Six White Dwarfs with Circumstellar Silicates',
Astronomical Journal, CXXXVII/2 (2009), pp. 3191–7.

20 A DIAMOND BORN FROM DEATH

1 A. Wolszczan and D. A. Frail, 'A Planetary System around the
Millisecond Pulsar PSR1257 + 12', *Nature*, 355 (1992), pp. 145–7.
2 A. Wolszczan, 'Confirmation of Earth-Mass Planets Orbiting the
Millisecond Pulsar PSR B1257+12', *Science*, 264 (1994), pp. 538–42.
3 P. Podsiadlowski, 'Planet Formation Scenarios', in *Planets
Around Pulsars*, ed. J. A. Philipps, S. E. Thorsett and S. R.
Kulkarni (San Francisco, CA, 1992), pp. 149–65.
4 B. Margalit and B. D. Metzger, 'Merger of a White Dwarf-
neutron Star Binary to 1029 Carat Diamonds: Origin of the
Pulsar Planets', *Monthly Notices of the Royal Astronomical Society*,
CDLXV/3 (2017), pp. 2790–803.

EPILOGUE: YET MORE WORLDS

1 T. Barclay et al., 'A Revised Exoplanet Yield from the Transiting
Exoplanet Survey Satellite (TESS)', *Astrophysical Journal
Supplement Series*, CCXXXIX/1 (2018), id. 2.
2 'TESS Planet Count and Papers', https://tess.mit.edu, accessed
2 February 2020.
3 V. B. Kostov et al., 'The L 98-59 System: Three Transiting,
Terrestrial-size Planets Orbiting a Nearby M Dwarf ',
Astrophysical Journal, CLVIII/1 (2019), id. 32.
4 J. G. Winters et al., 'Three Red Suns in the Sky: A Transiting,
Terrestrial Planet in a Triple M Dwarf System at 6.9 Parsecs',
article id.:1906.10147, arxiv.org, accessed 7 August 2019.
5 E. R. Newton, 'TESS Hunt for Young and Maturing Exoplanets
(THYME): A Planet in the 45 Myr Tucana–Horologium
Association', *Astrophysical Journal*, DCCCLXXX/1 (2019), id. L17.
6 European Space Agency, PLATO *Definition Study Report (Red
Book)* (2017).

APPENDIX: THE TWENTY WORLDS

1 M. Mayor and D. Queloz, 'A Jupiter-mass Companion to a Solar-type Star', *Nature*, 378 (1995), pp. 355–9.

2 G. W. Henry et al., 'A Transiting "51 Peg-like" Planet', *Astrophysical Journal*, 529 (2000), pp. L41–L44; T. Mazeh et al., 'The Spectroscopic Orbit of the Planetary Companion Transiting HD 209458', *Astrophysical Journal*, DXXXII/1 (2000), pp. L55–L58.

3 F. Bouchy et al., 'ELODIE Metallicity-biased Search for Transiting Hot Jupiters, II: A Very Hot Jupiter Transiting the Bright K Star HD 189733', *Astronomy and Astrophysics*, CDXLIV/1 (2005), pp. L15–L19.

4 L. Hebb et al., 'WASP-19b: The Shortest Period Transiting Exoplanet Yet Discovered', *Astrophysical Journal*, DCCVIII/1 (2010), pp. 224–31.

5 A. Pál et al., 'HAT-P-7b: An Extremely Hot Massive Planet Transiting a Bright Star in the Kepler Field', *Astrophysical Journal*, DCLXXX/2 (2008), pp. 1450–56.

6 J.-P. Beaulieu et al., 'Discovery of a Cool Planet of 5.5 Earth Masses through Gravitational Microlensing', *Nature*, 439 (2006), pp. 437–40.

7 M. J. Holman et al., 'Kepler-9: A System of Multiple Planets Transiting a Sun-like Star, Confirmed by Timing Variations', *Science*, 330 (2010), p. 51.

8 J. A. Carter et al., 'Kepler-36: A Pair of Planets with Neighboring Orbits and Dissimilar Densities', *Science*, 327 (2012), p. 556.

9 N. M. Batalha et al., 'KEPLER's First Rocky Planet: Kepler-10b', *Astrophysical Journal*, DCCXXIX/1 (2011), id. 27.

10 V. Rajpaul et al., 'Pinning Down the Mass of Kepler-10c: the Importance of Sampling and Model Comparison', *Monthly Notices of the Royal Astronomical Society*, CDLXXI/1 (2017), p. L125–L130.

11 F. Fressin et al., 'Kepler-10 c: A 2.2 Earth Radius Transiting Planet in a Multiple System', *Astrophysical Journal Supplement*, CXCVII/1 (2011), p. 5.

12 A.-M. Lagrange et al., 'A Probable Giant Planet Imaged in the β Pictoris Disk. VLT/NaCo deep L'-band imaging', *Astronomy and Astrophysics*, CDXCIII/2 (2009), pp. L21–L25.

13 C. Marois et al., 'Direct Imaging of Multiple Planets Orbiting the Star HR 8799', *Science*, 322 (2008), pp. 1348–52.

14 This is larger than the 6 Jupiter masses reported in M. C. Liu et al., 'The Extremely Red, Young L Dwarf PSO J318.5338-22.8603: A Free-floating Planetary-mass Analog to Directly Imaged Young Gas-giant Planets', *Astrophysical Journal Letters*, DCCLXXVII/2 (2013), id. L20, but takes into account the increased age reported in A. S. Binks and R. D. Jefferies, 'A Lithium Depletion Boundary Age of 21 Myr for the Beta Pictoris Moving Group', *Monthly Notices of the Royal Astronomical Society*, CDXXXVIII/1 (2014), pp. L11–L15.

15 Liu et al., 'The Extremely Red, Young L Dwarf PSO J318.5338-22.8603'.

16 K. L. Luhman, 'Discovery of a ~250 K Brown Dwarf at 2 pc from the Sun', *Astrophysical Journal*, DCCLXXXVI/2 (2014), id. L18.

17 M. Gillon et al., 'Temperate Earth-sized Planets Transiting a Nearby Ultracool Dwarf Star', *Nature*, 533 (2016), pp. 221–4.

18 M. Gillon et al., 'Seven Temperate Terrestrial Planets around the Nearby Ultracool Dwarf Star TRAPPIST-1', *Nature*, 542 (2017), pp. 456–60.

19 G. Anglada-Escudé et al., 'A Terrestrial Planet Candidate in a Temperate Orbit around Proxima Centauri', *Nature*, 536 (2016), pp. 437–40.

20 L. Hebb et al., 'WASP-12b: The Hottest Transiting Extrasolar Planet Yet Discovered', *Astrophysical Journal*, CDXCIII/2 (2009), pp. 1920–28.

21 A. Wolszczan, 'Confirmation of Earth-Mass Planets Orbiting the Millisecond Pulsar PSR B1257+12', *Science*, 264 (1994), pp. 538–42.

22 A. Wolszczan and D. A. Frail, 'A Planetary System around the Millisecond Pulsar PSR1257 + 12', *Nature*, 355 (1992), pp. 145–7.

23 NASA Goddard Spaceflight Center, 'Lunar and Planetary Science', https://nssdc.gsfc.nasa.gov, accessed 17 August 2019.

24 *The Extrasolar Planet Encyclopedia*, http://exoplanet.eu, accessed 17 August 2019.

25 Masses from S. L. Grimm et al., 'The Nature of the TRAPPIST-1 Exoplanets', *Astronomy and Astrophysics*, 613 (2018), id. A68; all other numerical values from L. Delrez et al., 'Early 2017 Observations of TRAPPIST-1 with Spitzer', *Monthly Notices of the Royal Astronomical Society*, CDLXXV/3 (2018), pp. 3577–97.

ACKNOWLEDGEMENTS

Thanks to Jim Geach for acting as the series editor for Universe, to Vera Schleich for her support and patience while I was writing the book, to Emma Rigby and Beth Biller for their helpful comments on the draft and to Ludmila Carone for helpful discussions on exoplanet atmospheres. Finally thank you also to the countless astronomers on whose papers and talks this book is based. Any errors are my own.

PHOTO ACKNOWLEDGEMENTS

The author and publishers wish to express their thanks to the following sources of illustrative material and/or permission to reproduce it.

ALMA (ESO/NAOJ/NRAO): p. 95; the author: pp. 13, 14, 21, 32, 39, 43, 50, 51, 58, 63, 70, 87; Davide De Martin (ESA/Hubble): p. 175; ESO/M. Kornmesser (cropped from original): p. 153; ESO/A.-M. Lagrange et al.: p. 104; Harvard-Smithsonian Center for Astrophysics/David Aguilar: p. 79; photo George Hodan: p. 160; N. Metcalfe & Pan- STARSS1 Science Consortium: p. 120 (arrow added by the author); NASA: pp. 42, 86; NASA, ESA, and G. Bacon (STSCI): pp. 162, 170; NASA, ESA and Allison Loll/Jeff Hester (Arizona State University): p. 175; NASA, ESA, C. R. O'Dell (Vanderbilt University), and M. Meixner, P. McCullough, and G. Bacon (Space Telescope Science Institute): p. 167; photos NASA/JPL-Caltech: pp. 25, 138–9, 177 – planet labels added by author; photos NASA/JPL-Caltech (adapted by the author from original Spitzer images): p. 129; reproduced by kind permission of *Private Eye* magazine and Banx: p. 83; Jason Wang (UC Berkeley/Caltech), Christian Marois (NRC Herzberg) & Keck Observatory – planet labels added by author: p. 109.

INDEX

Page numbers in *italics* refer to illustrations